Topics in Applied Physics
Volume 88

Available Online

Topics in Applied Physics is part of the SpringerLink service. For all customers with standing orders for Topics in Applied Physics we offer the full text in electronic form via SpringerLink free of charge. Please contact your librarian who can receive a password for free access to the full articles by registration at:

http://www.springerlink.com/orders/index.htm

If you do not have a standing order you can nevertheless browse through the table of contents of the volumes and the abstracts of each article at:

http://www.springerlink.com/series/tap/

There you will also find more information about the series.

Springer
Berlin
Heidelberg
New York
Hong Kong
London
Milan
Paris
Tokyo

Topics in Applied Physics

Topics in Applied Physics is a well-established series of review books, each of which presents a comprehensive survey of a selected topic within the broad area of applied physics. Edited and written by leading research scientists in the field concerned, each volume contains review contributions covering the various aspects of the topic. Together these provide an overview of the state of the art in the respective field, extending from an introduction to the subject right up to the frontiers of contemporary research.
Topics in Applied Physics is addressed to all scientists at universities and in industry who wish to obtain an overview and to keep abreast of advances in applied physics. The series also provides easy but comprehensive access to the fields for newcomers starting research.
Contributions are specially commissioned. The Managing Editors are open to any suggestions for topics coming from the community of applied physicists no matter what the field and encourage prospective editors to approach them with ideas.
See also: http://www.springer.de/phys/books/tap

Managing Editors

Dr. Claus E. Ascheron

Springer-Verlag Heidelberg
Topics in Applied Physics
Tiergartenstr. 17
69121 Heidelberg
Germany
ascheron@springer.de

Dr. Hans J. Koelsch

Springer-Verlag New York, Inc.
Topics in Applied Physics
175 Fifth Avenue
New York, NY 10010-7858
USA
hkoelsch@springer-ny.com

Assistant Editor

Dr. Werner Skolaut

Springer-Verlag Heidelberg
Topics in Applied Physics
Tiergartenstr. 17
69121 Heidelberg
Germany
skolaut@springer.de

Junji Tominaga   Din P. Tsai (Eds.)

Optical Nanotechnologies

The Manipulation of Surface and Local Plasmons

With 168 Figures

Springer

Prof. Junji Tominaga
National Institute
of Advanced Industrial Science
and Technology (AIST),
Laboratory for Advanced Optical Technology
(LAOTECH)
1-1-1 Higashi
305-8562 Tsukuba
Japan
j-tominaga@aist.go.jp

Prof. Din P. Tsai
National Taiwan University
Department of Physics
10617 Taipei
Taiwan
dptsai@phys.ntu.edu.tw

Cataloging in Publication Data applied for

Bibliographic information published by Die Deutsche Bibliothek.
Die Deutsche Bibliothek lists this publication in the Deutsche Nationalbibliografie; detailed bibliographic data is available in the Internet at <http://dnb.ddb.de>.

Physics and Astronomy Classification Scheme (PACS):
73.20.Mf, 42.25.Fx, 42,65.Pc, 42.65.-k, 42.79.Vb, 81.07.-b, 73.21.b, 73.22.-f

ISBN 978-3-642-07896-5 e-ISBN 978-3-540-45871-5
ISSN print edition: 0303-4216
ISSN electronic edition: 1437-0859

Springer-Verlag Berlin Heidelberg New York
a part of Springer Science+Business Media

http://www.springer.de

Softcover reprint of the hardcover 1st edition 2003

Cover design: *design & production* GmbH, Heidelberg

Preface

Surface and localized surface plasmons have been attractive research fields in physics and chemistry since their discovery because of their unique features. Although surface plasmons and related optics including near-field optics have a long history, unfortunately they have not been applied much to industry. The main issues to be overcome were the control and manipulation of the localized electromagnetic field confined in the nanometer region on the surface. The precise control and manipulation of the surface plasmons were not realized until scanning near-field optical microscope (SNOM) was invented in the 1980s. Since then, a huge amount of interesting data has been accumulated experimentally and theoretically. In particular, techniques to fabricate a nano-sized aperture at an end of an optical fiber have enabled the observation of optical images of less than 100 nm in combination with advanced SNOM technologies. In optical data storage, it is expected that the use of the optical near-field will enable the ultimate storage density in the near future.

This book focuses on the application and manipulation of optical near-field and surface plasmons. The body of this book is separated into two main parts: "super-RENS", which is a technique to overcome several issues in optical near-field data storage, and the basic concept for manipulating surface plasmons. The former part may orient towards optical data storage and its related new technologies including the special properties of thin films in dynamic conditions. The later part contains individual attractive approaches and characteristics in surface plasmons and localized surface plasmons. Super-RENS is the abbreviation of "super-resolution near-field structure", which was invented in 1998. This technology has the potential to enable optical near-field recording free from high-speed nanometer flight of a near-field optical head. Since the invention, several attractive features generated in multilayers consisting of super-RENS have been understood and their relationship with surface plasmons has been discussed. Readers who are not familiar with optical data storage technologies may not understand the contents well, but the concept of the system includes many hints for manipulating optical near-field and surface plasmons in each research field from physics to chemistry without SNOM systems.

Finally, we wish to thank Dr. Claus Ascheron for supporting our book project. We would like to especially acknowledge all the contributors to this project: Professors T. D. Milster, W.-C. Liu, U. C. Fischer, T. Fukaya, T. C. Chong, J.-J. Greffet, O. J. F. Martin, and Drs. J. H. Kim, T. Shima, T. Nakano, M. Kuwahara, L. P. Shi. Also, I would like to thank to the former members of LAOTECH, especially D. Büchel and C. Mihalcea who moved to Seagate Technology.

Tsukuba and Taipei, *Junji Tominaga*
January 2003 *Din Ping Tsai*

Contents

Coherent Spontaneous Emission of Light Due to Surface Waves
Jean-Jacques Greffet, Rémi Carminati, Karl Joulain,
Jean-Philippe Mulet, Carsten Henkel, Stéphane Mainguy,
and Yong Chen 163

The Super-Resolution Near-Field Structure and Its Applications

Junji Tominaga

Laboratory for Advanced Optical Technology (LAOTECH), National Institute of Advanced Industrial Science and Technology (AIST),
1-1-1 Higashi, Tsukuba, 305-8562, Japan
j-tominaga@aist.go.jp

Abstract. The super-resolution near-field structure (super-RENS) was first proposed in 1998 in order to overcome several crucial issues in optical near-field recording. Since the invention, a number of studies to understand the mechanism and near-field properties have been carried out. In this paper, the background of near-field optical recording methods proposed prior to super-RENS disks is reviewed and possible future approaches and applications are described.

1 The Optical Near-Field and Its Application for High-Density Data Storage

Since the invention of the scanning tunneling microscope, microscopy has evolved with respect to tunneling-electron, atomic-force, shear-force and surface-potential detection. The near-field scanning optical microscope (NSOM) was invented in the early 1980s by Pohl's laboratory in IBM Zurich [1]. Since this invention, a huge amount of papers have been published on the near-field observation of material surfaces and biochemical molecules, the detection of fluorescence from semiconductor lasers, the improvement of the resolution images and the challenge to optical data storage [2,3,4,5,6,7,8,9]. First, near-field optical recording (NFO recording) was carried out by *Betzig's* group at AT&T in the US, and a bit density as high as 45 Gbits/in^2 was achieved [8]. Although the recording density was very high in comparison with that in DVD discs and a 60-nm resolution was obtained, its scanning speed and recording area were limited by the ability of piezo-actuators moving a magneto-optical (MO) recording film sample. As shown in Fig. 1, the scanning speed of the system based on piezo-electric actuators were typically at most 100 µm, and the recording area was less than about $100 \times 100\ \mu m^2$. In order to overcome the crucial issues in NFO recording, flying heads with several small optical apertures have been designed as hard-disk flying sliders for the last 5 years [10,11]. However, the input laser energy is mostly wasted as reflected light, and at present only a few percent of energy is transferred into the near-field generation. *Terris* and his co-workers at IBM, alternatively, tried to introduce a solid immersion lens (SIL) into NFO recording [9,12,13].

J. Tominaga and D. P. Tsai (Eds.): Optical Nanotechnologies,
Topics Appl. Phys. **88**, 1–23 (2003)

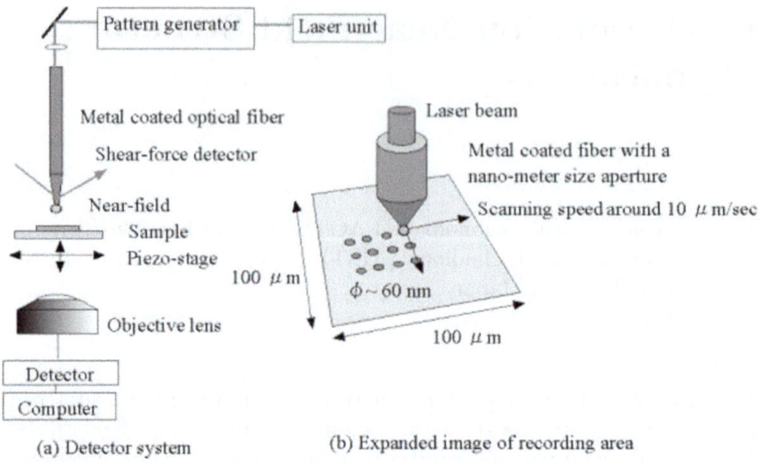

Fig. 1a,b. Typical near-field recording using an optical fiber probe with a nanometer-size aperture. The near-field aperture scans over the MO recording film and the polarization of the MO Kerr rotation is detected as a signal

Although SIL cannot overcome the diffraction limit, the lens numerical aperture (NA) dramatically improves by 2.0, which depends on the refractive index n of the lens material. Using a super-hemispherical lens for SIL recording, the resolution is further improved by n^2. Thanks to *Mansfield* and *Kino*, NFO-recording research and development using SIL has rapidly expanded worldwide since the 1990s (Fig. 2) [14,15]. In NFO recording using both a small aperture or a SIL, it should be noticed that the most basic problems all result in the high-speed space control technique between a recording medium and the flying aperture or the bottom of the SIL. The fundamental idea of the super-resolution near-field structure (super-RENS) was certainly based on removing the problem.

2 Basic Principle of the Near-Field Optical Readout

In order to understand the behavior of the NFO readout intuitively, let us assume a system consisting of two different screens in proximity and discuss the optical interaction between the two screens by using Fourier optics [16]. Hence, one screen has a small hole and the second screen has two holes, as shown in Fig. 3. First, we have to transform the shape of the step functions (hole) as Fourier functions. The step functions of the holes are then expressed as a function of $\sin(x)/x$. The electromagnetic field at the first screen with two slits is transformed into E_1:

$$E_1(k_x, z=0) = 4E_0 \cos k_x d \frac{\sin k_x L}{k_x} . \qquad (1)$$

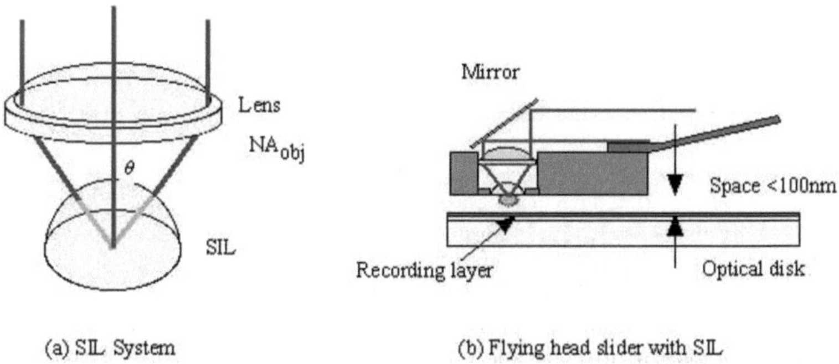

Fig. 2a,b. Solid immersion lens (SIL) system and the flying head with the SIL. In SIL, the effective NA can be increased up to the refractive index n of the SIL material. Thus, $NA_{effect} = NA_{obj} \cdot n$. (here, NA_{obj} is the objective lens NA). The high spatial frequency beyond NA/λ becomes an evanescent wave at the bottom surface of the SIL

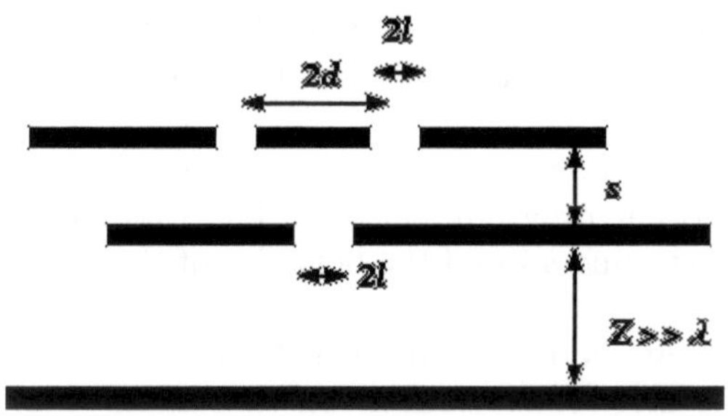

Fig. 3. Two-screen model for the Fourier treatment of near-field coupling

Hence, the second-screen position is taken as $z = 0$, and the distance between the two slits and the slit diameter are $2d$ and $2l$, respectively; E_0 is the incident electromagnetic field. Here, the propagated light field $E_\mathrm{d}(x, z = Z)$ towards the z-direction is described by

$$E_\mathrm{d}(x, z = Z) = \frac{1}{2\pi} \int_{-\infty}^{+\infty} \mathrm{d}k_x \exp(-k_x x) E_1(k_x, z = 0) \times \exp[-\mathrm{i}(k^2 - k_x^2)^{\frac{1}{2}} Z] \,. \quad (2)$$

In far-field optics and optical microscope system, Z is far longer than the optical wavelength and we need to consider only values of k_x from $-\omega/c$ to $+\omega/c$ because the $\exp[-\mathrm{i}(k^2 - k_x^2)^{1/2} Z]$ term beyond this range becomes real and is rapidly eliminated; the optical near-field is thus the contribution from the outer integration region. Now, separating the second screen from the first one with a slit by a distance $s (s < \lambda)$, the near-field integration term becomes large and is convoluted as the coupling field. Finally, we derive the following equation:

$$E_{\mathrm{d}}(x, z = Z) = \frac{1}{(2\pi)^2} \int_{-\omega/c}^{\omega/c} \mathrm{d}k_x \exp(-k_x x) \exp[-\mathrm{i}\,(k^2 - k_x^2)^{\frac{1}{2}}\,(Z - s)]$$
$$\times \int_{-\infty}^{+\infty} \mathrm{d}k'_x 4E_0 \cos k'_x d \frac{\sin(k'_x l)}{k_x} \exp\left[-\mathrm{i}\left(k^2 - k'^2_x\right)^{\frac{1}{2}} s\right]$$
$$\times 2 \cos k'_x d \frac{\sin(k_x - k'_x) l}{k_x} \exp[\mathrm{i}\,(k_x - k'_x)\,X]\,. \quad (3)$$

Here, k'_x is the wave vector along the second screen. The light passing through the first screen produces the field coupling and it is scattered by the second-screen slits. In the second term, the spatial frequencies beyond k_x give a contribution with a real exponential result which is rapidly reduced with increasing s; however, the contribution is now quite large because of small s less than λ. As s gets smaller, the contribution from $k > k_x$ becomes large. It means that as the near-field aperture is close to the surface by SNOM or other technical methods, the resolution is improved. This is the explanation and basic principle of the resolution improvement by using near-field optics (Fig. 4).

The simple computer simulation based on this principle is now easily programmed by "*Mathematica*™" or other program software. The simulation results are very simple but educative to understand the behavior of the near-field and its scattering.

In Figs. 5 to 8, two mark patterns are Fourier transformed and the second screen with a slit moves behind the first screen with a constant distance s. In the case of $s = 0$ (two screens are in contact), mostly the same pattern of the slit shape (or feature) are reproduced. As s increases, however, the edges of each slit are vague and the two edges are no longer obtained.

By comparing Figs. 5 and 6, we can understand the resolution characteristics of near-field readout. That is, the resolution mostly depends on the observing aperture size and the space s. In Fig. 5, the resolution is obtained as $s \approx 2\lambda$, but as $1/2\lambda$ in Fig. 6. Following the simulations, then, the distance of the slits in the first screen is reduced to the ratio of 1:1. The simulated results are more interesting (Figs. 7, 8). Now, the resolution of the two slits is no longer obtained, even at $1/10\ \lambda$. In order to improve the resolution, the slit size of the second screen has to be further reduced by 1/5, and the result is shown in Fig. 8. Therefore, it is understood that at least the aperture size should be a quarter or less in comparison with the slit size in near-field readout.

The Fourier analysis is very convenient and intuitively understandable, but it is of scalar calculation. This means that the analysis cannot apply to systems dealing with vector elements, for example, polarization and multi-scattering. The results therefore do not give the exact analytical solutions.

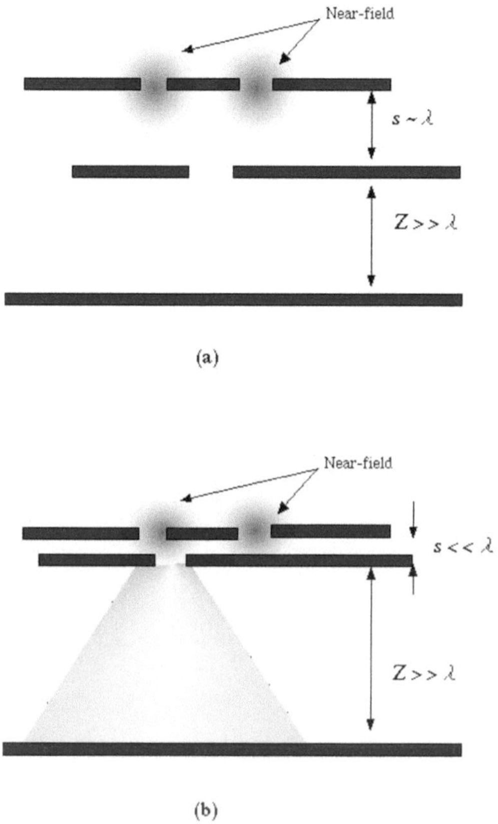

Fig. 4. Intuitive image of near-field coupling of two screens with small slits or apertures. (**a**) $s \approx \lambda$, the near-field cannot couple with the slit of the second screen. (**b**) $s \ll \lambda$, the near-field is scattered by the slit and propagated

3 Super-Resolution Near-Field Structure

In microscopy, a SIL lens is typically used to increase the resolution by filling the space between the lens and the surface of an object with oil. Because of the transparency of the oil, the near-field passes through it and therefore a higher-resolution image is obtained due to its high refractive index in comparison with air. "Why does nobody replace the air gap with a liquid or solid thin film for NFO recording?" This was the first hint for the super-RENS invention. Fortunately, a method of fabricating a small optical aperture in optical data storage and improving the resolution was reported using an optical phase change film (GeSbTe): this was called optical super-resolution (SR) [17]. In a combination of both methods, it was shown that high-speed NFO recording and readout would be possible by inserting a dielectric film between an optical recording layer and the mask layer. In super-RENS, an Sb thin film was adopted instead of GeSbTe because the near-field signal might be very weak and the background noise should be reduced as low as possible in comparison with the SR detection [18]. As-deposited films of GeSbTe,

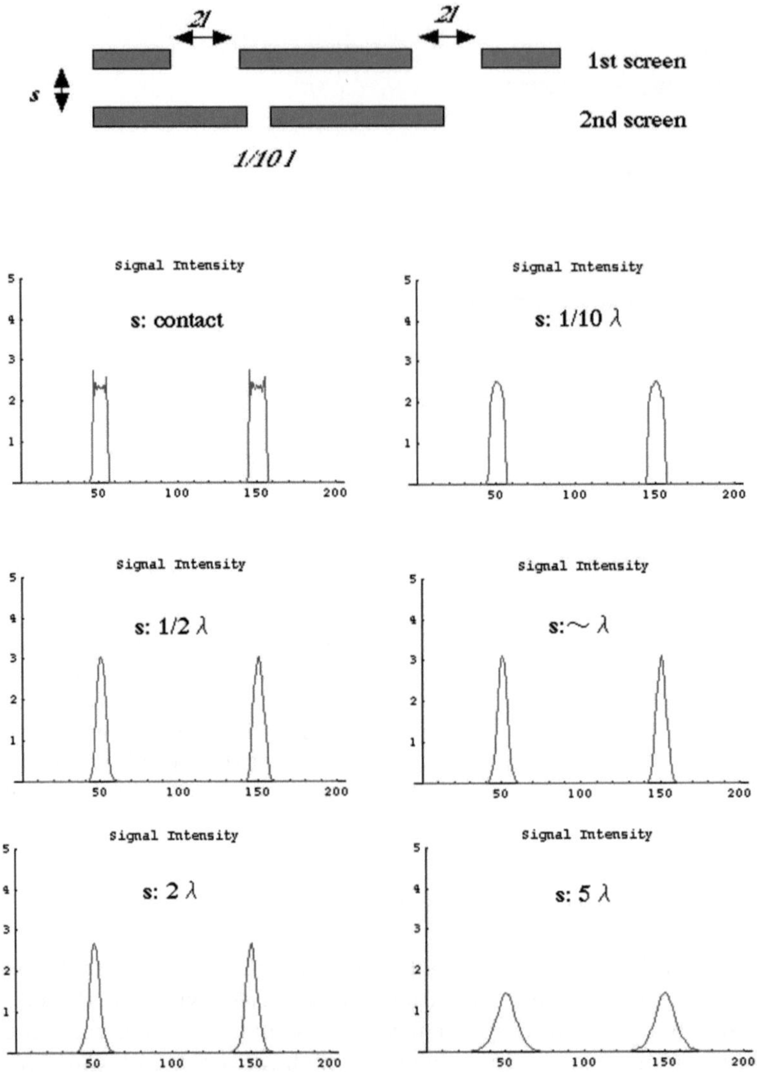

Fig. 5. Fourier simulation of two-screen problems. Here, the first screen has two slits and the distance is taken to the ratio of 4:1 (aperture). The slit size of the second screen is 1/5 of that of the first screen's aperture. The horizontal axis from this figure to Fig. 8 means the numbers of the simulation cell units

in general, have an amorphous state; therefore, the film has to be initially crystallized by radiating a high-power laser beam, which produces crystalline grains with random size. The grains might increase the background noise. As-deposited Sb films are, on the other hand, in a crystalline state; the grain size is controlled uniquely over a 12 cm disk substrate by adjusting the depo-

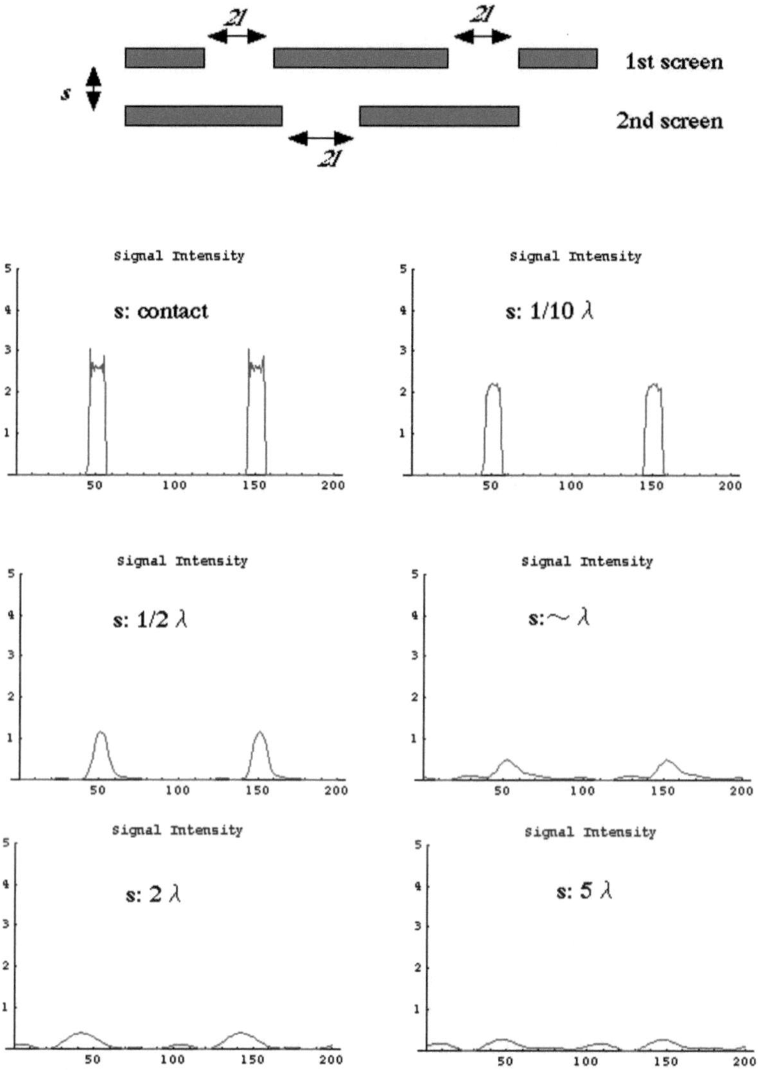

Fig. 6. Fourier simulation with a different aperture size. Here, the slit size of the second screen is expanded to the same size as the first screen's slit

sition conditions. The first super-RENS disk was designed for noise reduction (Fig. 9).

The control of the optical aperture in the Sb film (thickness 15 nm) is carried out by adjusting the power of a focused laser beam and the disk rotation speed. As the laser power increases, the light energy is absorbed in the Sb film and then the temperature increases close to the melting point. Increasing the rotation speed, on the other hand, reduces the temperature

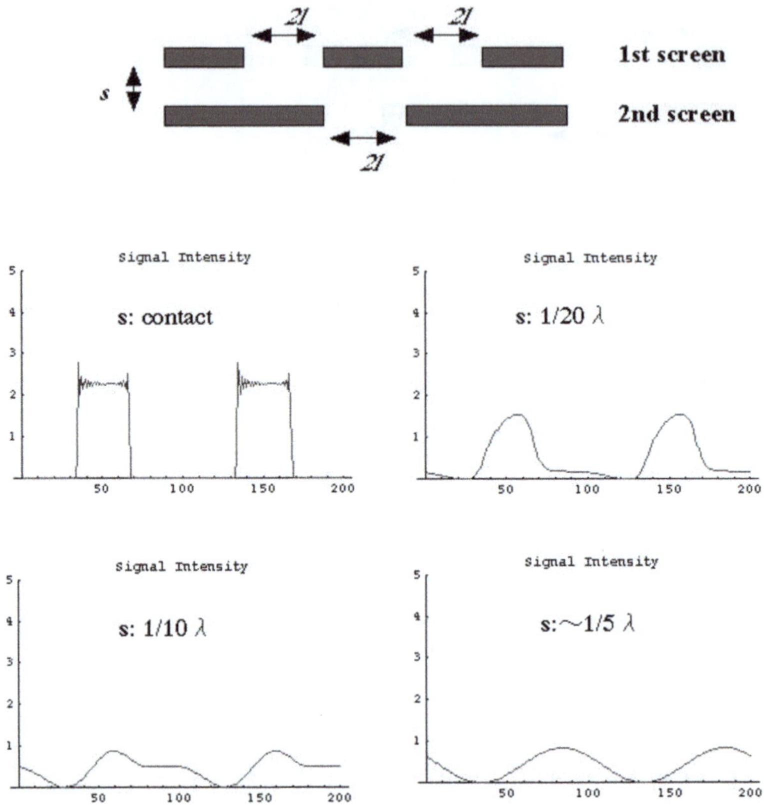

Fig. 7. Fourier simulation with the slit duty ratio of 1:1. Hence, the slit size of the second screen is the same as that of the first screen

because the absorbed light energy density per time is reduced by the speed. In the static condition, the Sb film cannot sustain the film condition and it gets smeared out or diffused in the multilayers because of the heat accumulation. This is the trick of optical disk technology in super-resolution. The first experiment of super-RENS was carried out in March 1998. Typical data for the resolution is shown in Figs. 10 and 11. Even using a 635-nm wavelength and a lens NA of 0.6, the resolution reaches 60 nm:1/10 λ [19].

The behavior of the near-field generated from an aperture in the Sb film is also confirmed by using a more accurate computer-simulation called the finite-differential time-domain (FDTD) method. Using the refractive indices of the crystalline and amorphous states of the Sb thin film: $3.36 + 5.55j$ and $4.51 + 3.66j$, the electromagnetic field in the TM-mode is evaluated and shown in Fig. 12. As shown in Fig. 12, the Sb layer partially blocks the incident light and the Gaussian profile of the beam is sharper behind the aperture.

The resolution limit of the super-RENS is of scientific interest, but the signal intensity is not high enough for engineering applications. In optical

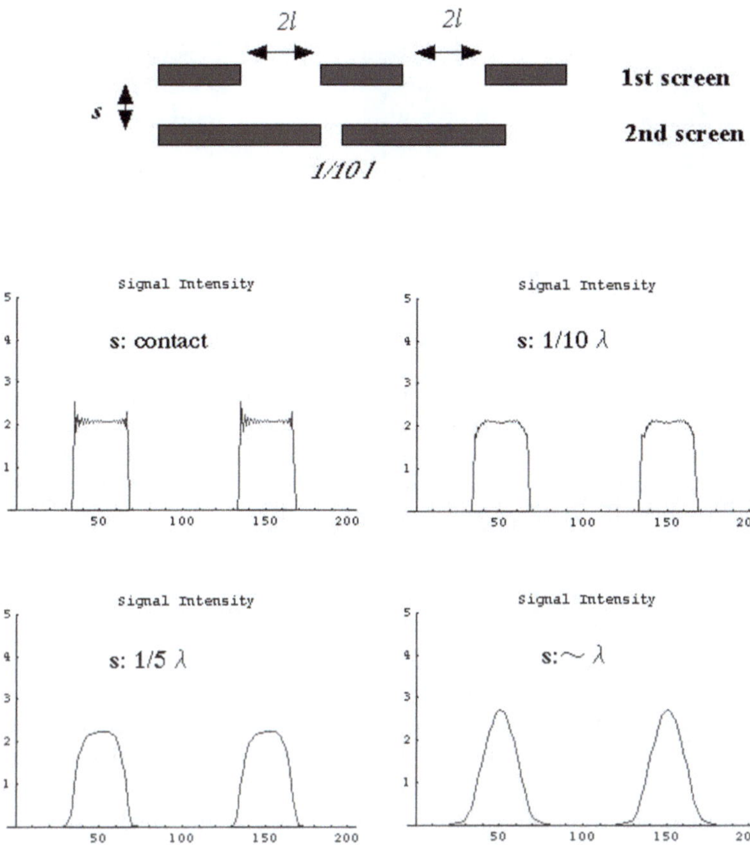

Fig. 8. Fourier simulation with the 1/5-size aperture of Fig. 7

data storage, a carrier-to-noise ratio (CNR) level of more than 40 dB is generally required to overcome laser noise, medium noise, detector sensitivity and so on. In our group, then we gradually turned our direction to fabricating and designing a new super-RENS disk using local or surface plasmon enhancement. According to a number of papers related to surface and local plamsons, the plasmon effect and enhancement has more potential than using evanescent wave.

One of the easy ways of fabricating the new super-RENS disk was to replace the Sb layer with a silver layer. However, little enhanced signals were observed except for a little super-ROM effect, which was reported by a TDK group, although this effect was explained by surface plasmons generating over several marks in a beam spot [21]. Fortunately, it was noticed that silver oxide (AgOx) decomposition might generate surface plasmons, according to the author's old memory of inorganic compact-disc recordable (CD-R) research [22,23]. AgOx thin film is easily deposited by RF magnetron sputtering

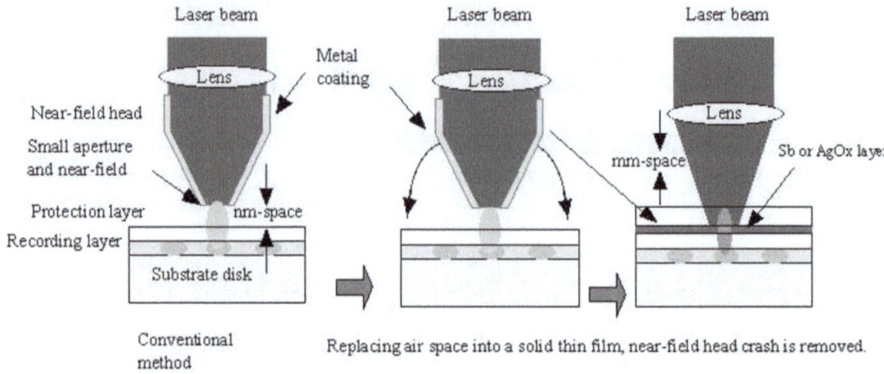

Fig. 9. The concept of super-resolution near-field structure (super-RENS). A metal coating over the NSOM fiber head is spread over the recording medium. Focusing the laser beam on the metal coating, an optical aperture is generated and eliminated by the laser power. The metal coating has to be thin and have optical nonlinear characteristics. In super-RENS, Sb or silver oxide is used

with a gas mixture of Ar and oxygen, and the composition is well-controlled by the gas ratio. Focusing a laser beam in the film, AgOx is rapidly decomposed into Ag and oxygen to produce Ag clusters. If the cluster formation occurs in the super-RENS disk, it may generate local plasmons and enhance signals by the strong light scattering. The newly designed super-RENS disk using the Ag_2O composition was fabricated and rotated on a DVD disk tester. At more than 2.5 mW (3.5 mW for the case of the Sb film), very strong scattered signals with more than 30 dB were detected at 200 nm mark patterns (Fig. 13) [24,25].

4 Recording Mark Trains and Surface Plasmons in Optical Data Storage

The new super-RENS disk characteristics are very interesting and helpful for understanding the plamon properties. Figures 14 and 15 give us attractive hints hidden in ultra-density optical data storage in future.

The super-RENS disk was specially designed for the studies of the plasmon interaction with the electromagnetic field generated around the phase change marks. In the disk, the second AgOx layer was deposited and the distance d from the recording film was changed from 20, 30, 40, 80 to 100 nm [26]. For d greater than 80 nm, the scattered light intensities are almost constant. The intensities of marks beyond the diffraction limit depend on d and increase, while those of marks in the far-field detection do not. It clearly means that smaller marks beyond the diffraction limit scatter more incident light than the far-field signals and this effect results in some electromagnetic field interactions between the marks and the scattering center generated in the

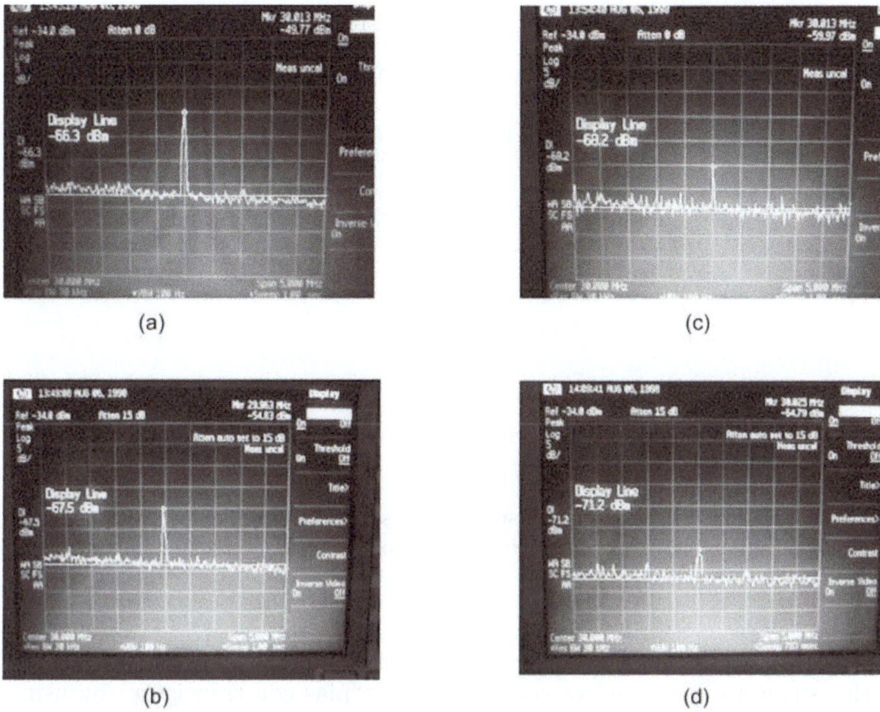

Fig. 10. Recorded small mark signals observed from the super-RENS disk using an Sb thin film. The disk was rotated on a DVD drive tester with 635 nm λ and NA 0.6 (the resolution limit is approximately 270 nm). (**a**) CNR of 100 nm marks (recording power: 6.0 mW, readout power: 3.1 mW), (**b**) CNR of 80 nm marks (recording power: 5.0 mW, readout power: 3.2 mW), (**c**) CNR of 70 nm marks (recording power: 5.0 mW, readout power: 3.1 mW), and (**d**) CNR of 60 nm marks (recording power: 4.6 mW, readout power: 2.7 mW). The disk rotation speed for each recording was adjusted to the best condition between 1.8 m/s and 3.0 m/s, and the readout was carried out between 3.6 m/s and 6.0 m/s

AgOx film. In turn, more energy is accumulated over the smaller marks in the super-RENS disk. It is supposed that surface plasmons are generated over the small marks in the focused laser spot, and the plasmons are scattered by the scattering center in the proximity. Thus, the marks in the phase change film may play the role of a grating for the surface plasmons [27,28]. Because of increasing wave vector k_x along the mark trains by the grating-pitch effect, the total wave vector k becomes imaginary [$k = (k_z^2 - k_x^2)^{1/2}$ and $k_x > k_z$, where k_z is the normal wave vector to the disk surface]. As the mark size becomes smaller and smaller, and once the size is beyond the diffraction limit, the reflected beams with higher-order diffraction no longer travel away from the disk surface as far-field light, but get trapped as surface plasmons (Fig. 16). This is very interesting. The recorded marks in the super-RENS

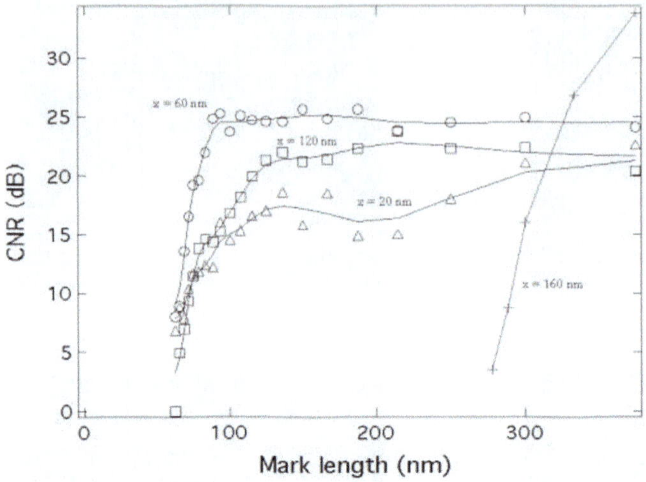

Fig. 11. Relationship between the signal intensity (CNR) and mark resolution by super-RENS disk. X means the intermediate protection layer thickness between the Sb and GeSbTe layers [20]

disk work as something like a photonic device accumulating surface plasmons. If this suggestion is true, we can fabricate a "plasmon transistor" by using two laser beams crossing the spots in the super-RENS disk [29]!

5 Local Plasmon Photonic Transistor and Scattering

The system configuration of the plasmon transistor and the concept are depicted in Figs. 17 and 18. Crossing two laser beams at a mark-pattern recorded track, the first laser beam is used to accumulate the surface plasmons over the marks and the second laser is used to generate the light-scattering center in the AgOx film. The second laser may also generate the surface plasmons. The power of the first laser is modulated at frequencies (around 15 to 20 MHz), which are detected by a photodetector installed in the red laser pickup unit. Thus, the transmitted signal passing through the disk is observed. Here, the second laser is not modulated but the power is increased or decreased; therefore, the signal detected by the photodetector is the purely transmitted signal through the disk (Fig. 18). In the actual experiment, a blue laser (405 nm) unit is used for the first laser, and a red laser (635 nm) unit for the second one. Figure 19 shows the relationship between the signal gain and red laser power, when the mark size is varied. In the cases of 300 nm and 500 nm marks, the signal gains change a little. However, the gains for the smaller marks of 200 nm and 150 nm beyond the diffraction limit are remarkable and the gain is more than 10 dB. Without recording marks, the gain was mostly 0 dB. It clearly means that the power of the red

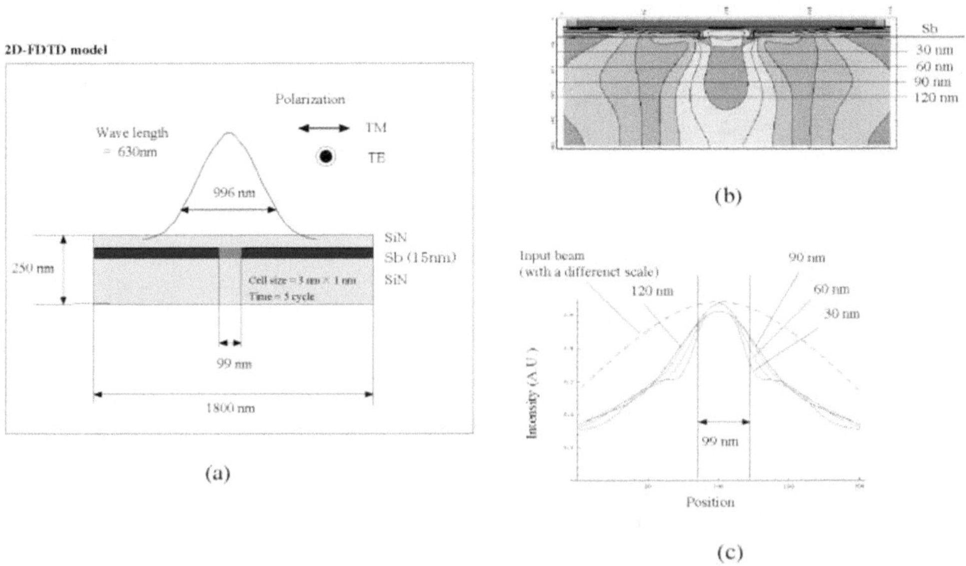

Fig. 12. (**a**) The simulation model of the near-field generation behind an aperture in the Sb film. The aperture size is 99 nm, (**b**) the electromagnetic field of the TM-mode and (**c**) the intensity profiles at different positions behind the aperture [20]

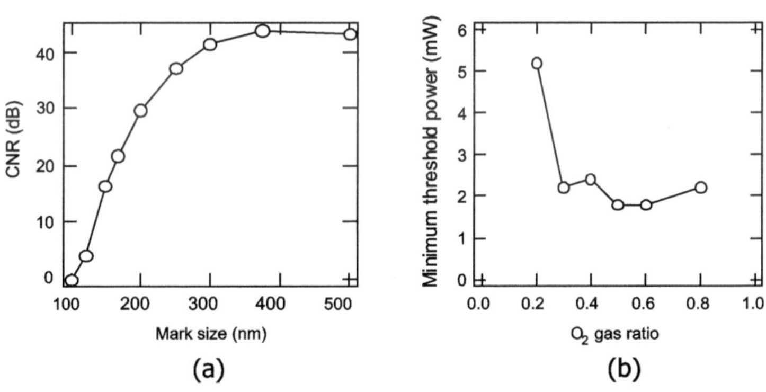

Fig. 13. The resolution of a AgOx super-RENS disk. (**a**) The resolution curve against the recorded marks and (**b**) the minimum readout power to generate the super-RENS effect against the AgOx deposition condition to the total gas-flow ratio ($Ar+O_2$). The Ag-rich phase is produced with a ratio between 0.0 and 0.3, and the Ag_2O-rich phase with around 0.4 to 0.6. Finally, the AgO-rich phase is at more than 0.6

laser beam can control the blue transmitted light by the surface plasmons accumulated over the marks.

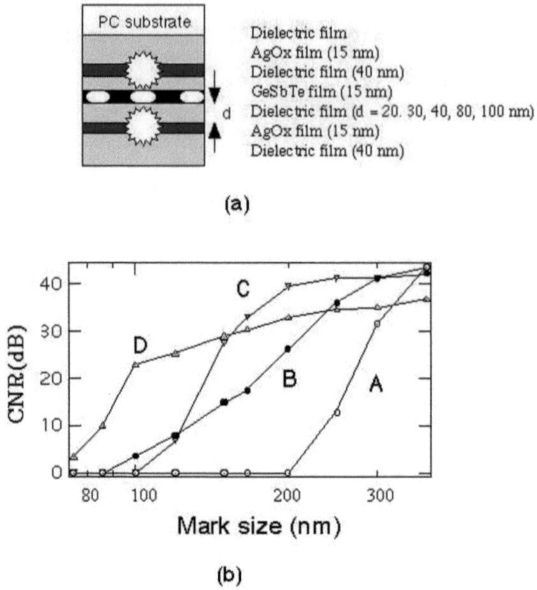

Fig. 14. Resolution dependence of new super-RENS disks using double-AgOx layers. (**a**) The structure of the double-AgOx super-RENS disk and (**b**) the resolution limit of recorded marks depending on d. A: Far-field readout from the original super-RENS disk using the single layer, B: the near-field readout of A, C: near-field readout from the double super-RENS disk with $d = 30\,\text{nm}$ distance, and D: the disk with $d = 20\,\text{nm}$. As the second AgOx layer moves closer to the recording layer, the resolution is improved. Hence, the laser wavelength is 635 nm and the NA is 0.6

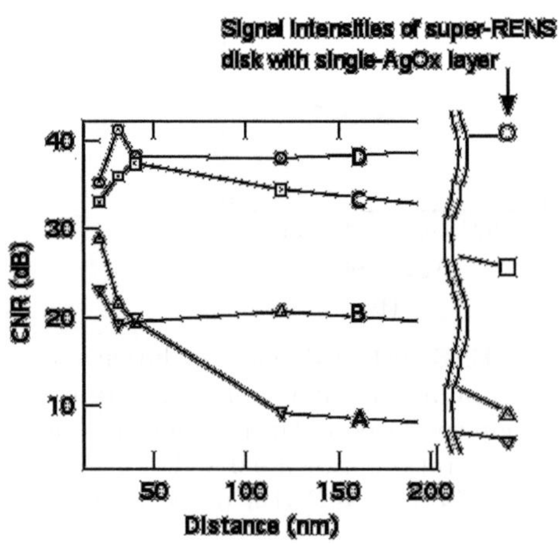

Fig. 15. The near-field signal intensity dependence of the intermediate layer thickness d. A: the mark size is 100 nm, B: 150 nm, C: 200 nm, and D: 400 nm. Hence, the laser wavelength is 635 nm and the NA is 0.6

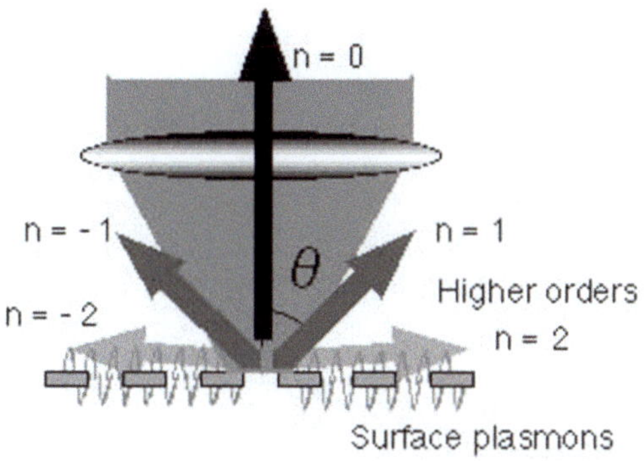

Fig. 16. Schematic images of surface plasmon generation over small marks. Higher-order diffraction beams become surface plasmons by reducing the mark size and pitch. On reducing the pitch a, the wave vector along the mark trains increases and then finally the total wave vector becomes imaginary

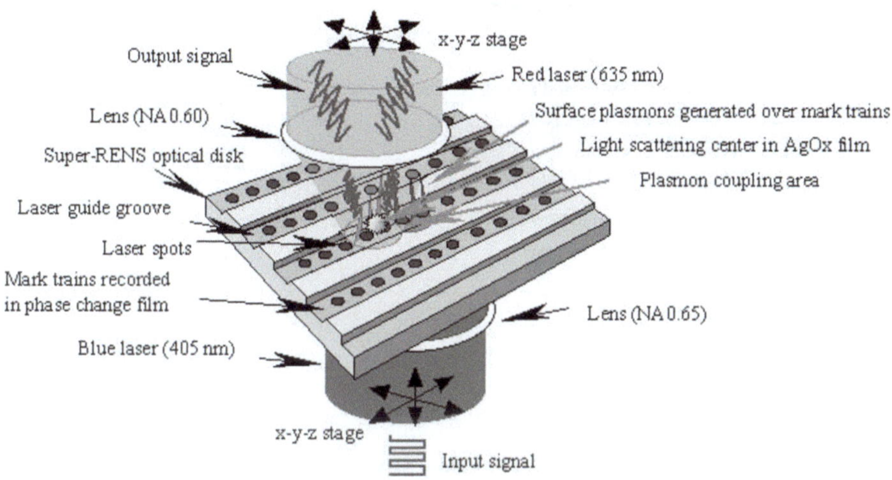

Fig. 17. The schematic picture of a plasmon transistor. Crossing two laser beams at one mark recorded track, part of the incident light energy is accumulated as surface plasmons, and then scattered by the light-scattering center generated in the AgOx layer

It can be verified that high-order diffraction plays a role in the surface plasmon generation. By displacing one of the laser beam spots forward and backward relatively, the high-order diffracted signals can be detected in the system (Fig. 18). When the position of the red laser beam spot was shifted to that of the blue one along the disk grooves, the transmitted blue laser intensity showed a Gaussian profile and the shape was not changed by increasing the red laser power, generating the scattering center in the AgOx layer (Fig. 20). When the red spot was shifted by jumping the tracks, the same peak curve was obtained, too. Hence, the track width was 1.2 μm and the groove depth was 60 nm. No mark was recorded in the disk. The red laser pickup lens was placed above the disk surface at about 2.0 mm. The results are very reasonable because of no diffraction points. After recording

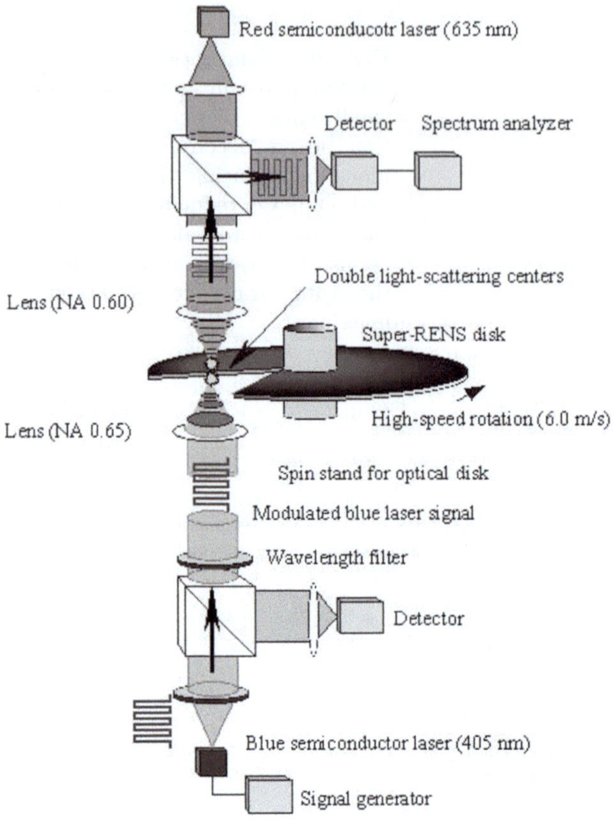

Fig. 18. Experimental setup for the plasmon transistor. The system consists of two laser beam optical heads with 405 nm (NA 0.65) and 635 nm (NA 0.6) wavelengths. The focused points are semi-automatically controlled by focus-servo and tracking-servo systems. The transmitted light of blue laser from the super-RENS disk is detected by a photodetector installed in the red optical pickup

marks with different sizes, 300 nm, 200 nm and 150 nm in the phase-change film by the red laser, the transmitted blue intensities changed as in Fig. 21. Interestingly, the intensity curves are separated with one peak to two, and the intensity from the 150 nm mark-pattern recorded track is much higher than in the other cases. In addition, on reducing the mark size, the separated peaks gradually merged. According to the results, it is clear that the red laser lens detected the high-order diffracted beams from the mark patterns and much higher-order diffraction may propagate along the mark patterns as surface plasmons. In conclusion, it was confirmed that the device actually works by the plasmons scattering between the surface plasmons accumulated along the mark trains and the scattering center generated in the AgOx layer.

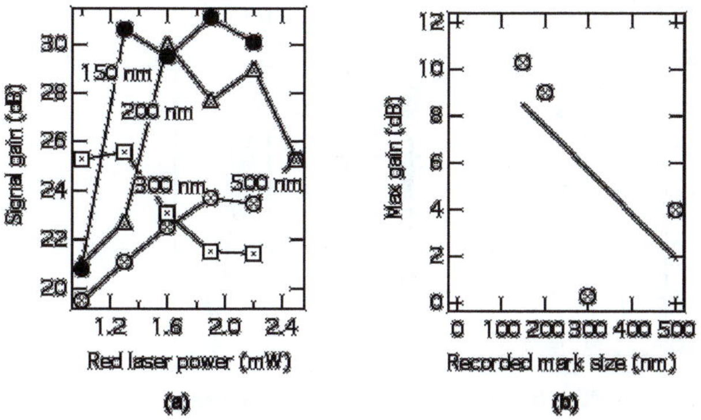

Fig. 19. The signal gain of the plasmon transistor. (**a**) The relationship between the signal gain and the red laser power and (**b**) the gain versus the recorded mark size

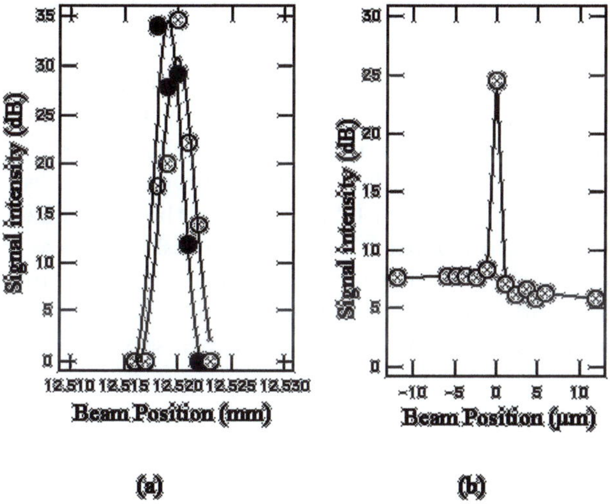

Fig. 20. The blue laser signal intensity change versus the red laser position shift: (**a**) the red laser was shifted along the groove, and (**b**) across the groove. Hence, the blue laser was power-moduated at 14.8 MHz with the power between 0.5 and 0.7 mW. *Open circles* show the intensity at the red laser power of 1.0 mW and *solid circles* at 2.0 mW, generating the scattering center

It could be understood that optical phase-change recorded marks in the crystalline and amorphous states accumulate surface plasmons when the size is less than the resolution limit: half the diffraction limit of the optical system, and the plasmons are scattered by the light-scattering center generated in the AgOx film. Then, what is going on when two light-scattering centers are in close proximity? The same sort of structure as in Fig. 14a without the recording layer was fabricated and the scattering from the two local plasmons

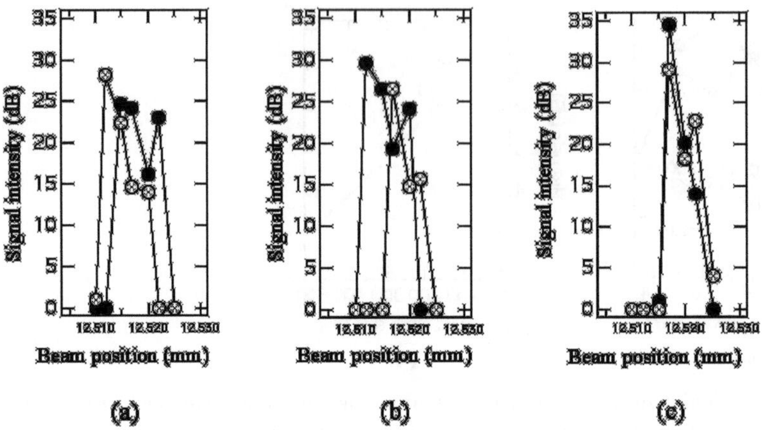

Fig. 21. The blue laser intensity changes after recording marks: (**a**) recorded mark pattern with 300 nm, (**b**) 200 nm and (**c**) 150 nm. The marks were recorded by the red laser at 7.0 mW, and the readout conditions were the same as those in Fig. 20

was examined. Prior to the experiment, an FDTD computer simulation was carried out and the results are shown in Fig. 22 and Fig. 23b. On decreasing the separation layer thickness d from 200 nm to 20 nm, the evanescent intensity in the layer is increased. On the other hand, the propagating electromagnetic field (E_x) shows a single peak at 0.25 to 0.30 of the normalized thickness to the wavelength (405 nm). Hence, the scattering disk (Ag) size is assumed to be 120 nm. It means that the space may induce optical cavity resonance because the refractive index of the separation layer is 2.2: the actual thickness is just half of the wavelength. The experimentally observed results

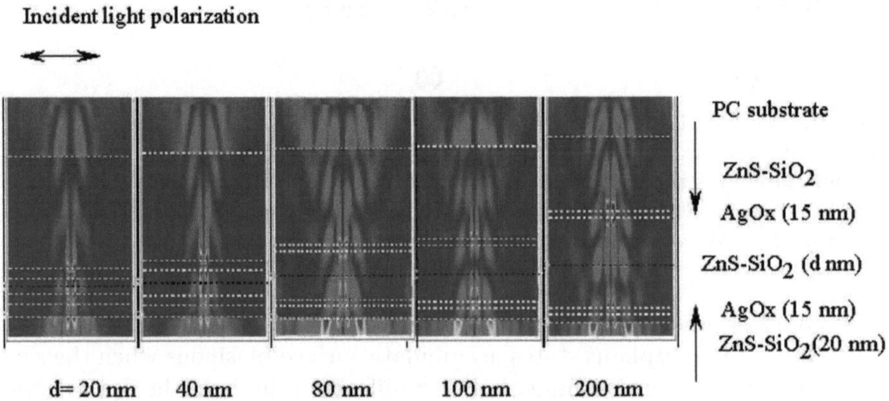

Fig. 22. The computer-simulated evanescent electromagnetic field in super-RENS disks with two light-scattering centers. Here, the third ZnS–SiO$_2$ layer thickness d is changed from 20 nm to 200 nm

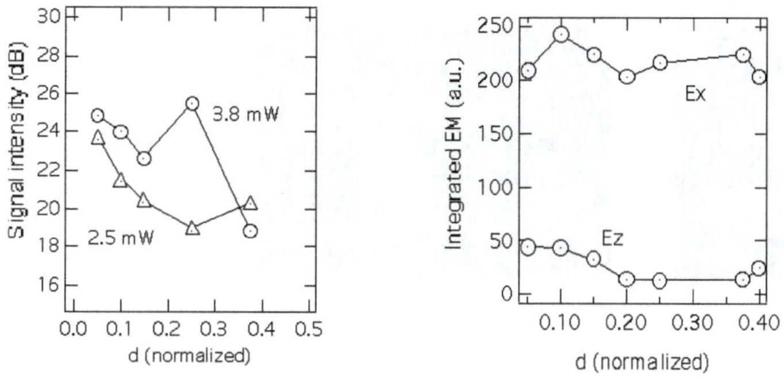

(a) (b)

Fig. 23. (**a**) Experimentally observed signal intensity change for different incident powers 2.5 mW and 3.8 mW. (**b**) Integrated electromagnetic field intensities of the evanescent (E_z) and propagated field (E_x)

are more interesting. The first signal enhancement occurred at more than 2.5 mW (red laser power) and the second enhancement was also observed at more than 3.5 mW (Fig. 23a). The intensity curve obtained at 2.5 mW seems to depend on the local plasmon coupling between two AgOx layers because of no resonance cavity due to a standing wave condition for the wavelength (405 nm) at normalized $d = 0.25$. On the other hand, the result observed at 3.8 mW seems to include the cavity effect.

Therefore, it was concluded that the super-RENS disk can be used as an optical switch by the control of the incident laser power, although the plasmon switch rotates in order to release the heat accumulation (Fig. 24).

6 Potential of Super-RENS in Future

As describing the characteristics of the super-RENS disks, the super-RENS seems to open a new research field in local and surface plasmons besides ultra-high density optical data storage. Very recently, our group members, *Büchel* and *Mihalcea*, detected very strong surface-enhanced Raman signals using sputtered AgOx films on glass or Si substrates. The Raman signal enhancement will be able to be applied to chemical or biochemical sensors in future [30,31]. Also, the plasmon transistor will be demonstrated not in the dynamic state like the rotating disk condition, but in a static state. We have recently designed a Si substrate with 100-nm pit patterns for the experiment (Fig. 25). The author very much looks forward to seeing such new devices based on plasmon scattering in the near future.

Acknowledgements

I wish to thank to all the members of LAOTECH, and collaborated members from TDK, Sharp, Pioneer, Minolta, JVC and Pulstec Corporations.

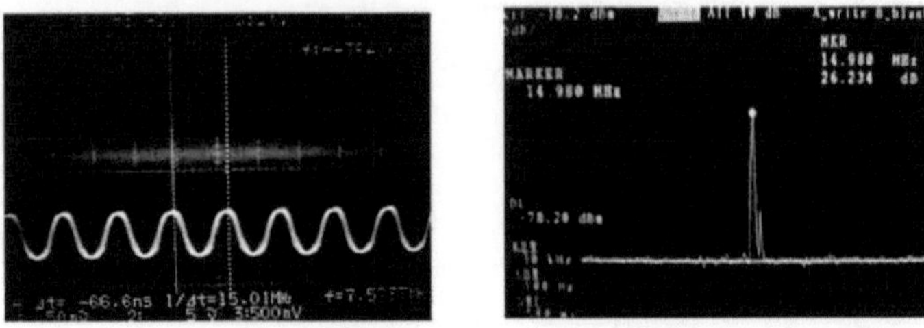

(a) $P^{Red}_{readout}$ = 1.5 mW

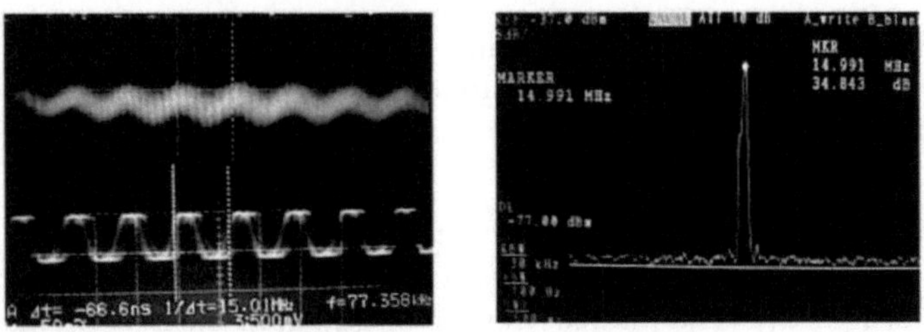

(b) $P^{Red}_{readout}$ = 3.5 mW

Fig. 24. Optical switching effect by the super-RENS disk with two light-scattering centers. The intermediate layer thickness is 20 nm, and the incident blue (405 nm) laser power is power-modulated between 4.5 mW and 1.0 mW at 15 MHz. The transmitted signal from the disk at the red laser power of 3.5 mW is 34.8 dB, and 26.2 dB at 1.5 mW. The *left-hand photos* of (**a**) and (**b**) show the transmitted signals (15 MHz) (*top*) and incident signals (*bottom*). The *right-hand photos* show the transmitted signals detected by the spectrum analyzer

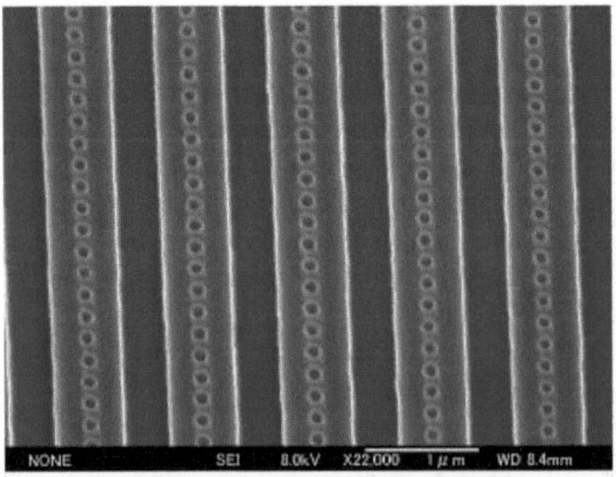

Fig. 25. SEM image of 100-nm pit patterns fabricated by the electron beam technique for the plasmon transistor

References

1. D.W. Pohl, W. Denk, M. Lanz: Appl. Phys. Lett. **44**, 651–653 (1984) 1
2. E. Betzig, R.J. Chichester: Science **262**, 1422–1425 (1993) 1
3. X.S. Xie, R.C. Dunn: Science **265**, 361–364 (1994) 1
4. H.F. Hess, E. Betzig, T.D. Harris, L.N. Pfeiffer, K.W. West: Science **264**, 1740–1745 (1994) 1
5. M. Isaacson, J.A. Cline, H. Barshatzky: J. Vac. Sci. Technol. B **9**, 3103–3107 (1991) 1
6. D.W. Pohl: In C. Seppard, T. Mulvey (Ed.): *Advances in Optical and Electron Microscopy*, (Academic Press, London 1991) 1
7. E. Betzig, J.K. Trautman, R. Wolfe, E.M. Gyorgy, P.L. Finn, M.H. Kryder, C.H. Chang: Appl. Phys. Lett. **61**, 142–144 (1992) 1
8. E. Betzig, S.G. Grubb, R.J. Chichester, D.J. DiGiovanni, J.S. Weiner: Appl. Phys. Lett. **63**, 3550–3552 (1993) 1
9. B.D. Terris, H.J. Mamin, D. Rugar, W.R. Studenmund, G.S. Kino: Appl. Phys. Lett. **65**, 388–390 (1994) 1
10. H. Ukita, Y. Katagiri, H. Nakada: Proc. SPIE **1449**, 248 (1991) 1
11. H. Yoshikawa, T. Ohkubo, K. Fukuzawa, L. Bouet, M. Yamamoto: Appl. Opt. **38**, 863–868 (1999) 1
12. B.D. Terris, H.J. Mamin, D. Rugar: Appl. Phys. Lett. **68**, 141–143 (1996) 1
13. I. Ichimura, S. Hayashi, G.S. Kino: Appl. Opt. **36**, 4339 (1997) 1
14. S.M. Mansfield, G.S. Kino: Appl. Phys. Lett. **57**, 2615–2617 (1990) 2
15. W. Yeh, M. Mansuripur: Appl. Opt. **39**, 302–315 (2000) 2
16. J.M. Vigoureux, F. Depasse, G. Girard: Appl. Opt. **31**, 3036–3045 (1992) 2
17. Y. Kasami, K. Yasuda, M. Ono, A. Fukumoto, M. Kaneko: Jpn. J. Appl. Phys. **35**, 423–428 (1996) 5
18. J. Tominaga, T. Nakano, N. Atoda: Appl. Phys. Lett. **73**, 2078–2080 (1998) 5
19. J. Tominaga, T. Nakano, T. Fukaya, N. Atoda, H. Fuji, A. Sato: Jpn. J. Appl. Phys. **38**, 4089–4093 (1999) 8
20. A. Sato, J. Tominaga, T. Nakano, H. Fuji, N. Atoda: Proc. SPIE **3864**, 157–159 (1999) 12, 13
21. T. Kikukawa, T. Kato, H. Shingai, H. Utsunomiya: Jpn. J. Appl. Phys. **40**, 1624–1628 (2001) 9
22. J. Tominaga, S. Haratani, K. Uchiyama, S. Takayama: Jpn. J. Appl. Phys. **31**, 2757–2759 (1992) 9
23. S. Haratani, J. Tominaga, H. Dohi, S. Takayama: J. Appl. Phys. **76**, 1297–1300 (1994) 9
24. J. Tominaga, H. Fuji, A. Sato, T. Nakano, N. Atoda: Jpn. J. Appl. Phys. **39**, 957–961 (2000) 10
25. J. Tominaga, D. Büchel, T. Nakano, T. Fukaya, N. Atoda, H. Fuji: Proc. SPIE **4081**, 86–94 (2000) 10
26. J. Tominaga, J.H. Kim, D. Büchel, L. Men, H. Fukuda, T. Nakano, T. Fukaya, N. Atoda, H. Fuji, T. Kikukawa, A. Sato, A. Tachibana, Y. Yamakawa, M. Kumagai: Jpn. J. Appl. Phys. **40**, 1831–1834 (2001) 10
27. H. Raether: *Surface Plasmons* (Springer, Hamburg 1988) 11
28. S. Kawata (Ed.): Near-Field Optics and Surface Plasmons Polaritons, Topics Appl. Phys. **81** (Springer, Heidelberg, Berlin 2001) 11

29. J. Tominaga, C. Mihalcea, D. Büchel, H. Fukuda, T. Nakano, N. Atoda, H. Fuji, T. Kikukawa: Appl. Phys. Lett. **78**, 2417–2420 (2001) 12
30. C. Mihalcea, D. Büchel, N. Atoda, J. Tominaga: J. Am. Chem. Soc. **123**, 7172–7173 (2001) 19
31. D. Büchel, C. Mihalcea, T. Fukaya, N. Atoda, J. Tominaga: Appl. Phys. Lett. **79**, 620–622 (2001) 19

Near-Field Optical Properties of Super-Resolution Near-Field Structures

Din Ping Tsai

Department of Physics, National Taiwan University,
Taipei 10617, Taiwan
dptsai@phys.ntu.edu.tw

Abstract. Super-resolution near-field structures, glass/SiN(170 nm)/Sb(15 nm) /SiN(20 nm) and glass/ZnS–SiO_2 (20 nm)/AgO_x (15 nm)/ZnS–SiO_2 (20 nm), have been studied by a tapping-mode tuning-fork near-field scanning optical microscope in transmission mode. Both propagating and evanescent field intensities were found at the focused spots of the surface of the super-resolution near-field structure. Images of the near-field intensity gradients at different excited laser powers showed that the area of the static evanescent intensity could be stably controlled. The enhancement of the near-field intensity, and the reduction of the focused spot through super-resolution near-field structure, glass/SiN(170 nm)/Sb(15 nm)/SiN(20 nm) or glass/ZnS–SiO_2 (20 nm)/AgO_x (15 nm)/ZnS–SiO_2 (20 nm) have been directly observed in the near field. The near-field interactions of the 15 nm Sb and AgO_x Layers have been investigated, respectively, and the localized surface plasmons excited at the focused laser spot were shown to be the key sources of the strong enhancement in the near field.

1 Introduction

Near-field optical recording was first proposed and demonstrated by *Betzig* et al. [1] on multilayered Pt/Co magneto-optical thin films using a near-field scanning optical microscope (NSOM). The most important advantage for near-field optical recording is the superior spatial resolution with no diffraction limit. To realize near-field optical recording in practice, *Terris* et al. [2] used a solid immersion lens (SIL) to reduce the mechanical damage caused by the near-field optical fiber probe and to achieve a higher recording speed. However, in the past few years, the control of the near-field distance between the SIL recording head and the recording medium, and the near-field aperture size of the SIL, are the major hurdles for commercial applications. Recently, *Tominaga* et al. [3,4,5,6] suggested and demonstrated that a 15 nm Sb thin film on the top of the phase-change (PC) recording layer (GeSbTe) within the near-field distance can produce similar results to a local near-field SIL. They have successfully shown that a multilayered structure of polycarbonate/SiN (170 nm)/Sb (15 nm)/SiN (20 nm)/GeSbTe (15 nm)/SiN (20 nm) on a digital versatile disk (DVD) gives the estimated recorded marks of 90 nm at a constant linear velocity of 2.0 m/s. The carrier-to-noise ratio (CNR) can be more than 10 dB. They named this multilayered structure, SiN (170 nm)/Sb

J. Tominaga and D. P. Tsai (Eds.): Optical Nanotechnologies,
Topics Appl. Phys. **88**, 23–34 (2003)

(15 nm)/SiN (20 nm), a super-resolution near-field structure (super-RENS), and considered the 15 nm Sb thin film as a nonlinear optical layer which controls the near-field optical aperture. Because the near-field distance can be easily controlled by the fixed spacing layer (20 nm SiN) between the nonlinear optical layer (15 nm Sb) and the recording layer (15 nm GeSbTe), the super-RENS is considered a more feasible way of achieving near-field optical recording with a simpler recording head design, less mechanical damage, and higher recording speed. Most recently, *Tominaga* et al. [7,8] showed that a new type of super-RENS using a 15 nm AgO_x thin film as the nonlinear optical layer, has a much stronger near-field intensity and better CNR. Their results also showed the laser powers for the recording and readout are 1 mW less than that used in the Sb type super-RENS.

The near-field optical interaction of either the 15 nm Sb or AgO_x thin film is obviously the key subject of the super-RENS. The working mechanism of the super-RENS is definitely an important foundation for various potential applications of the super-RENS. In particular, the function of the 15 nm Sb or AgO_x thin film at the focused laser spot is a very interesting issue for the super-RENS. Although *Fukaya* et al. [9] and *Ho* et al. [10] have studied the optical switching property of a light-induced pinhole or scatter in the Sb or AgO_x thin film, respectively, the measurement of either the Sb or the AgO_x type of super-RENS using a near-field scanning optical microscope (NSOM) is certainly a vital experiment for the understanding of their near-field optical properties [11,12,13,14]. In this paper, we report the study of the focused laser spot through a super-RENS sample in the transmission mode by using a tapping-mode NSOM. The direct near-field optical imaging of the transmitted laser spots through these nonlinear optical layers provides important information, and reveals the working mechanism of the super-RENS.

2 Tapping-Mode Tuning-Fork Near-Field Scanning Optical Microscopy

A tapping-mode tuning-fork near-field scanning optical microscope (TMTF-NSOM) [15,16] was used in our direct measurements of the transmission properties of the super-RENS. A schematic of the experimental setup is shown in Fig. 1. An inverted TMTF-NSOM was used to probe the near-field intensity at the surface of the sample in the transmission mode. The wavelength of the laser diode was 650 nm. The input laser power to the 40× objective was attenuated to 6 μW to prevent damage to the sample. The height of the near-field fiber probe was modulated at the tapping frequency (15–33 kHz) of the TMTF-NSOM. The feedback control of the near-field optical fiber probe was provided by the internal lock-in amplifier of a commercial electronic control unit of an atomic force microscope (AFM) [17], which is described elsewhere [15,18]. Optical radiation coupled to the tapered fiber tip operat-

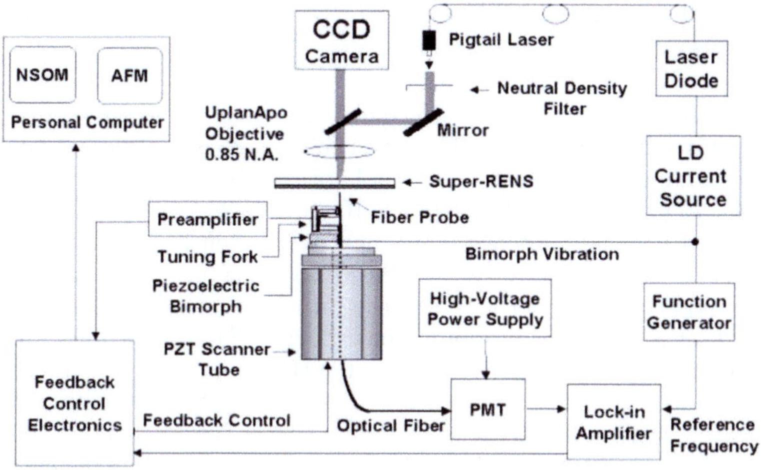

Fig. 1. Experimental setup of the tapping-mode tuning-fork near-field scanning optical microscope (TMTF-NSOM) in the transmission mode

ing in the near field was detected by a photomultiplier tube. The measured signal was fed to two external lock-in amplifiers which acquired signals at the modulation frequency ω and at 2ω and provided their ratio $I(2\omega)/I(\omega)$. The imaging of the measurements of the ratio of $I(2\omega)/I(\omega)$ gives the image of the local near-field intensity gradients [18]. Images of the near-field intensity, the topographic AFM image, and the near-field intensity gradients can be measured simultaneously.

The information obtained from the imaging of ratio measurements of $I(2\omega)/I(\omega)$ can be understood by considering that the transmitting intensity I through the substrate/sample/air may contain at least two terms. One is a propagating intensity term which is a constant I_c, and the other is an evanescent intensity term $I_e = I_0 e^{-2qz}$, i.e.

$$I = I_c + I_e = I_c + I_0 e^{-2qz} . \tag{1}$$

The gradient along the vertical z direction is dI/dz, and (1) leads to

$$dI/dz = -2qI_0 e^{-2qz} = -2qI_e . \tag{2}$$

Equations (1) and (2) show that for the transmitting intensity I including two terms $(I_c + I_e)$, the gradient of intensity is $-2qI_e$, which gives the contrast of q, which is the decay coefficient of the local evanescent field. The direct imaging contrast of the $I(2\,\omega)/I(\omega)$ ratio in our experiments is equivalent to the measured q contrast in (1) and (2) [18,19].

3 Near-Field Imaging of the Sb-Type Super-Resolution Near-Field Structure

Figure 2 shows the schematic of the near-field measurement in the transmission mode of our TMTF-NSOM. The lower left image is the optical micrograph of the tuning-fork with the near-field optical fiber probe under the focused sample. Figure 3a–c is the 10 μm × 10 μm images of AFM topography, transmitted near-field intensity, and near-field intensity gradients of a clean cover glass slip, respectively. The AFM image (Fig. 3a) shows the topographic features on the cover-glass surface. In Fig. 3b, the near-field intensity distribution of the focusing spot at the exiting surface of the glass slip shows an 860 nm full-width at half maximum (FWHM) spot size. The image of the near-field intensity gradients shown in Fig. 3c displays no feature at the same detecting sensitivity, because the transmitting intensity along the propagating axis (z) is a constant I_c, and the near-field intensity gradients ($\mathrm{d}I_c/\mathrm{d}z$) are zero over all the measured area. For the super-RENS sample of cover-glass/SiN(170 nm)/Sb(15 nm)/SiN(20 nm), the images of AFM topography, transmitted near-field intensity, and near-field intensity gradients are shown in Fig. 3d–f, respectively.

The AFM image (Fig. 3d) shows the topography of the surface of the SiN (20 nm) thin film. In Fig. 3e, the near-field intensity distribution of the focusing spot at the exiting surface of the glass/SiN (170 nm)/Sb (15 nm) /SiN (20 nm) shows a 780 nm FWHM spot size. The reduction of the spot size resulted from the higher index of refraction of SiN/Sb/SiN. The image of

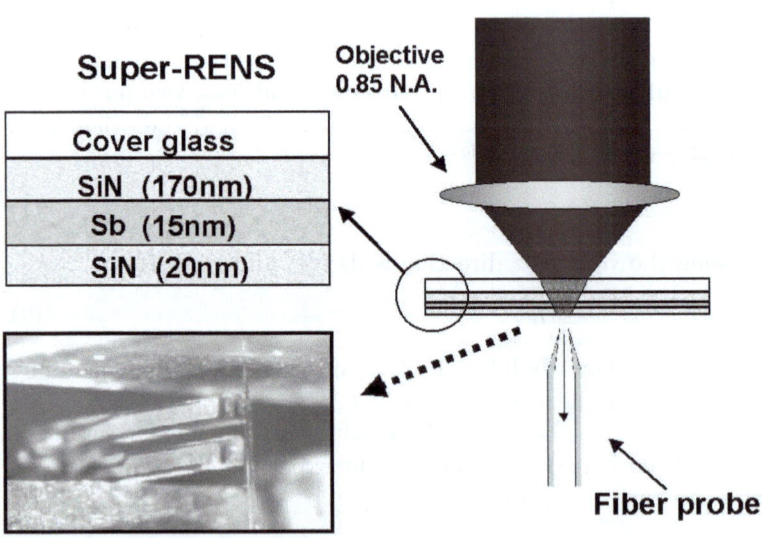

Fig. 2. Schemes of probing near-field distributions of the Sb-type super-RENS sample in the transmission mode

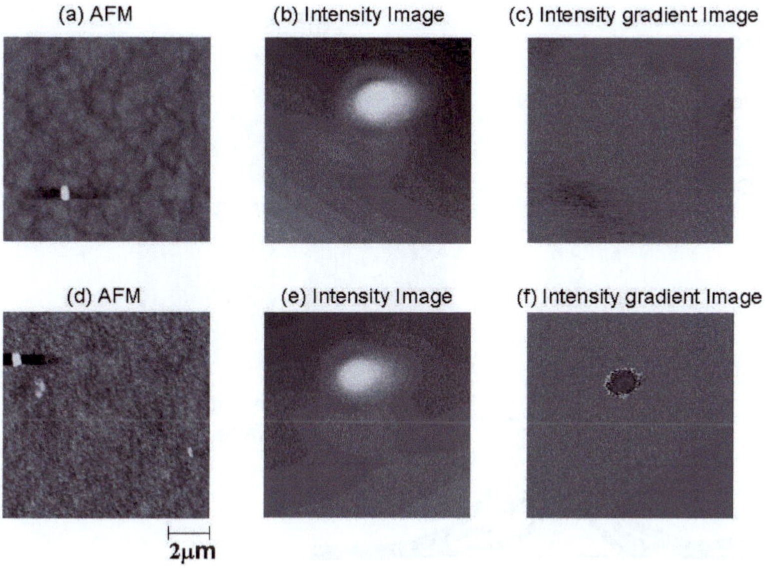

Fig. 3. (**a**)–(**c**) are images of AFM topography, transmitted near-field intensity, and near-field intensity gradients of a cover-glass slip, respectively. (**d**)–(**f**) are images of AFM topography, transmitted near-field intensity, and near-field intensity gradients of a Sb-type super-RENS sample, respectively

the near-field intensity gradients shown in Fig. 3f displays an 850 nm spot size with a measured negative ratio of $I(2\omega)/I(\omega)$ inside the spot, and relatively large and fluctuating positive values around the edge of the spot. This is important experimental evidence that the transmitted intensity I through the glass/SiN (170 nm)/Sb (15 nm)/SiN (20 nm) sample contains at least two terms. We believe that the evanescent field intensity shown in this case is due to the excitation of localized surface plasmons [20,21,22] by the focused laser beam at the Sb/SiN interfaces. The decay constant q can be regarded as the wave vector of the localized surface plasmons perpendicular to the interface of Sb/SiN. This can be further understood by the analysis of the measured near-field intensity profiles. Figure 4 shows the measured transmission near-field intensity profiles at a fixed incident laser power for the clean cover-glass slip and the super-RENS sample displayed in Fig. 3. Two calculated profiles are also shown in Fig. 4; one is the calculated near-field transmission intensity profile of the super-RENS sample, which is based on the near-field transmission profile of the clean glass slip, and the optical attenuation of the 15 nm Sb thin film; the other profile is the previously calculated profile multiplied by 2.25. The results of Fig. 4 clearly show that an enhancement of the near-field intensity profile has happened, and the enhancement in the center of the focused spot is stronger. The magnitude of this enhancement agrees with the enhancement of the surface plasmon effect [20,21,22].

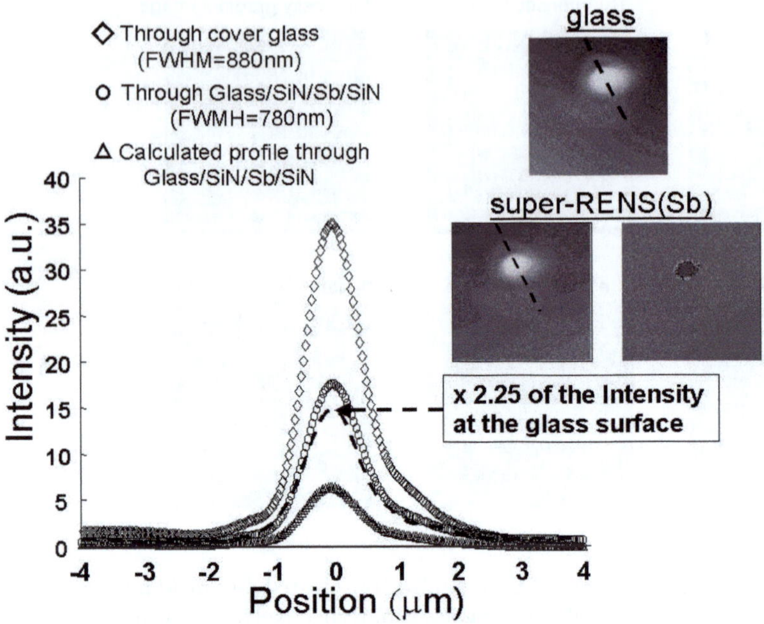

Fig. 4. Two measured near-field intensity profiles of the focused spot at the surfaces of the clean glass slip and Sb-type super-RENS sample are shown. A calculated transmission profile of the focused spot at the surface of the super-RENS sample, and its profile multiplied by 2.25, are also shown

Figure 5 shows the transmission images of the near-field intensity (upper row) and intensity gradients (lower row) at different incident laser intensities (0.42–2.43 μW). The results show that a focused laser can excite the super-RENS structure glass/SiN (170 nm)/Sb (15 nm)/SiN (20 nm) locally. The localized excitation area can be smaller than the laser-focused area. The area of the static evanescent intensity can be established in a very stable manner, and can be controlled by the focused laser power. The laser-excited area had large positive fluctuations locally, and they occurred around the edge of the static evanescent intensity area. We believe the formation of this evanescent intensity zone resulted from an ensemble of the localized surface plasmons that were excited and controlled by the focused laser, and the surface plasmon enhanced local evanescent field acts as the near-field "aperture" described by *Tominaga* et al. [3,4,5,6]. The near-field recording and reading of the super-RENS are the local photo-thermal interactions between this excited evanescent area and the 15 nm GeSbTe recording layer. One problematic issue shown in the near-field gradient images of Fig. 5 is the large positive fluctuations around the edge of the static evanescent intensity zone. This may be caused by the imperfection of the localized excitation at the SiN/Sb interface. The accuracy and perfection of the near-field recording marks in-

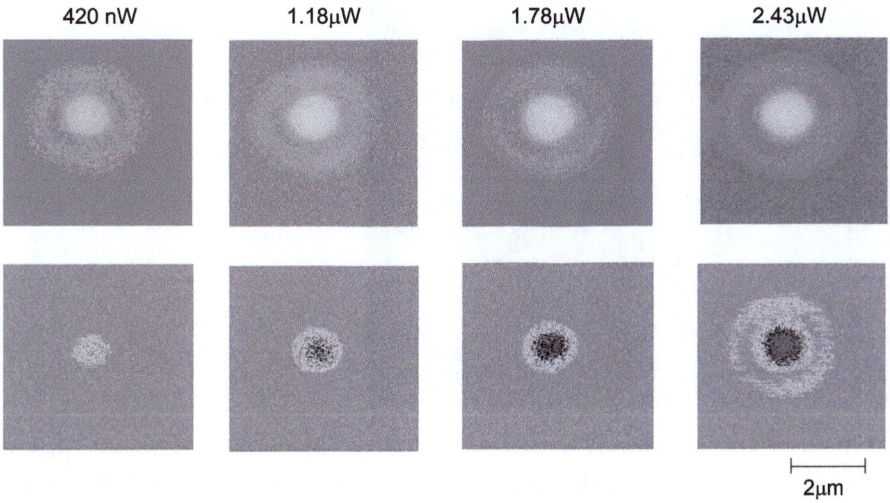

Fig. 5. 10 μm × 10 μm scanning images of the near-field intensity (*upper row*) and near-field intensity gradients (*lower row*) at the surface of the super-RENS sample, glass/SiN (170 nm)/ Sb (15 nm)/SiN (20 nm). The images were detected at the incident laser powers of 0.42, 1.18, 1.78, and 2.43 μW, respectively

duced by the super-RENS may be limited by this imperfect edge. The CNR of the super-RENS near-field optical disk may be significantly affected by this aperture-edge issue.

4 Near-Field Imaging of the AgO_x-Type Super-Resolution Near-Field Structure

Direct near-field imaging measurements of the transmission of a focused spot on the AgO_x-type super-RENS sample, glass/ZnS–SiO_2 (20 nm)/AgO_x (15 nm)/ZnS–SiO_2 (20 nm), were performed using a TMTF-NSOM in a similar manner to that described in previous sections. The nonlinear optical properties and near-field interactions of the AgO_x layer certainly can provide interesting and vital information about the working mechanism of the 15 nm AgO_x thin film.

The AgO_x super-RENS sample, cover-glass/ZnS–SiO_2 (20 nm)/AgO_x (15 nm)/ZnS–SiO_2 (20 nm), was fabricated by the reactive RF magnetron sputtering method. The dielectric layer, ZnS–SiO_2, has an index of refraction of about 2.25 + 0.01j, and the AgO_x has a refractive index of about 2.8 + 0.08j. Figure 6 is the TEM micrograph of the AgO_x-type super-RENS. Nano-composites of the AgO_x layer are clearly shown.

Figure 7 shows the near-field images of the intensity (upper row) and intensity gradients (lower row) of the glass/ZnS–SiO_2 (20 nm)/AgO_x (15 nm) /ZnS–SiO_2 (20 nm) sample at different incident laser powers. The wavelength

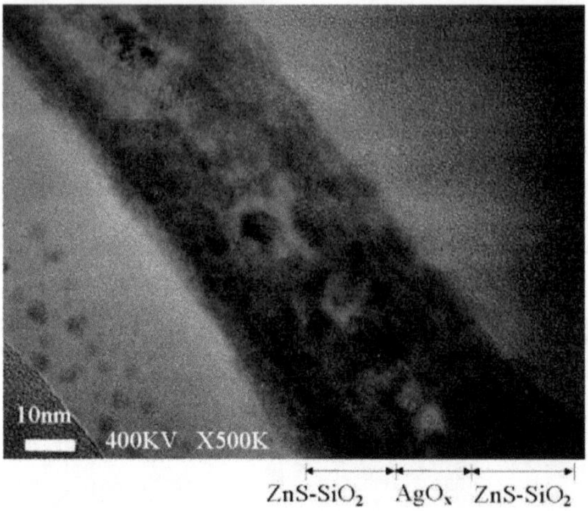

Fig. 6. TEM micrograph of the cross-section of the AgO_x-type super-RENS sample. The center layer consists of $ZnS–SiO_2$ (20 nm)/AgO_x (15 nm) /$ZnS–SiO_2$ (20 nm)

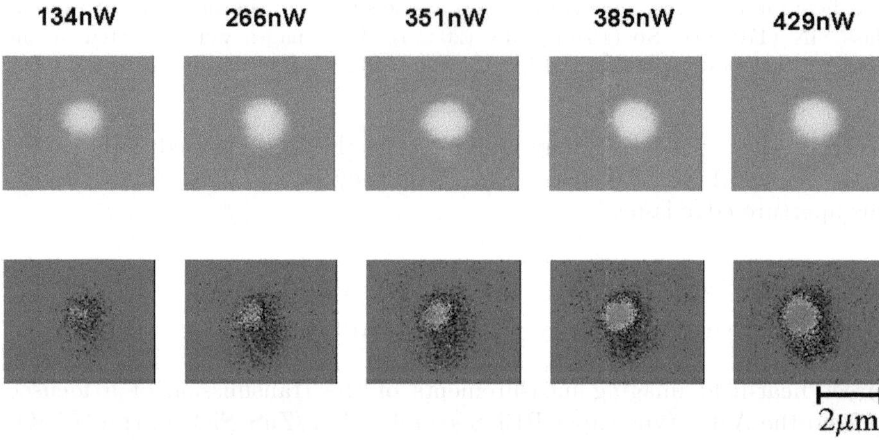

Fig. 7. Near-field images of the intensity (*upper row*) and intensity gradients (*lower row*) of the AgO_x super-RENS sample, glass/$ZnS–SiO_2$ (20 nm)/AgO_x (15 nm)/$ZnS–SiO_2$ (20 nm), at different incident laser powers

of the non-polarized laser source used in the experiments was 650 nm, and the input laser power was attenuated to 6 μW before the focusing objective shown in Fig. 1 to prevent the damage to the sample.

The near-field intensity distribution of the focusing spot of the glass/$ZnS–SiO_2$ (20 nm)/AgO_x (15 nm)/$ZnS–SiO_2$ (20 nm) sample showed a 780 nm FWHM spot size. The slightly reduction of the spot size resulted from the high effective index of refraction of the $ZnS–SiO_2$/AgO_x/$ZnS–SiO_2$ structure. The transmission images of the intensity gradients at different incident laser intensities (134–429 nW) show that a focused laser can locally excite the the glass/ZnS–SiO_2 (20 nm)/AgO_x (15 nm) /ZnS–SiO_2 (20 nm) sample, and the excited area can be smaller than the laser-focused area. The area of the static evanescent intensity can be established in a very stable manner, and can be controlled by the focused laser power as well. The situation is quite similar to that of the Sb-type super-RENS sample. However, the transmission images of the intensity gradients shown in Fig. 7 are different from that of Fig. 5. The laser-excited area of the AgO_x sample has a very strong evanescent field locally in comparison with that of the Sb-type sample. Large positive fluctuations occurred locally around the edge as well. This may be due to the strong scattering of the silver nano-clusters decomposed from the AgO_x. The localized surface plasmon can be excited by the focused laser spot, and can generate strongly scattered light in both propagating and evanescent modes [23,24]. *Liu* et al. [13] have demonstrated that strong near-field interactions can occur in both the rough interfaces and the existence of the embedded silver particles in the AgO_x thin film. The results of their finite-difference time-domain (FDTD) simulations show that the formation of the evanescent intensity zone resulted from an ensemble of the localized surface plasmons excited and controlled by the focused laser. The collective effect of this local strong scattering may be regarded as the near-field optical "aperture" mentioned by *Tominaga* et al. [3,4,5,6,7,8,9].

Measurements of the transmission near-field intensity profiles of the AgO_x sample also indicate the strong surface plasmon effect. Figure 8 shows the measured transmission near-field intensity profiles at a fixed incident laser power for a clean cover glass slip and a glass/ZnS–SiO_2 (20 nm)/AgO_x (15 nm)/ZnS–SiO_2 (20 nm) sample. The peak intensity of the transmission near-field profile of the AgO_x sample is strongly enhanced. The magnitude of the amplification is around two. The enhanced local electromagnetic field is due to localized surface plasmons of the decomposed Ag clusters, which is

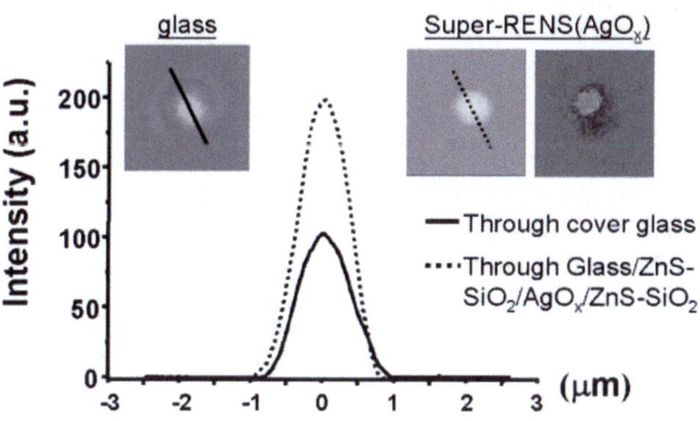

Fig. 8. Near-field intensity profiles of a cover-glass slip and a AgO_x super-RENS sample glass/ZnS–SiO_2 (20 nm)/AgO_x (15 nm)/ZnS–SiO_2 (20 nm)

not propagating but evanescent at the local metal–dielectric boundary. The amplitude of the locally enhanced field decays exponentially away from the interface, so the strong enhancement happened in the near-field region only. This is the main reason that the AgO_x-type super-RENS has better performance in near-field recording, and less laser power is needed for the readout and recording. The TEM micrograph shown in Fig. 6 also clearly indicates that the nano-composites of AgO_x play an important role in the uniform 15 nm thin film of AgO_x.

5 Summary

In summary, the near-field optical nonlinear properties of the SiN/Sb/SiN or ZnS–SiO_2/AgO_x/ZnS–SiO_2 thin films of the super-RENS have been observed experimentally using a TMTF-NSOM. Direct imaging of the near-field intensity and intensity gradients revealed that the intensity of the focused spot through the glass/SiN (170 nm)/Sb (15 nm)/SiN (20 nm) or glass/ZnS–SiO_2 (20 nm)/AgO_x (15 nm)/ZnS–SiO_2 (20 nm) sample has both propagating and evanescent terms. The near-field imaging results of the focused spot sizes of the static evanescent field intensity and their changes controlled by the excited laser power have successfully demonstrated the working mechanism of both types of the super-RENS. The ensemble of the localized surface plasmons and their photo-thermal energy transfer are the key factors of the near-field recording of the super-RENS. The strong enhancement that resulted from the localized surface plasmon excitation are a novel property of the super-RENS. The utilization of the local amplification of the near-field intensity may create many novel applications such as nano-photonic transistor and nano-optical computing devices. We believe that the development of the super-RENS opens up a brand new territory for nano-photonic research, and the applications of the nano-photonics of the localized surface plasmons have a great potential and a fascinating future.

References

1. E. Betzig, J. Trautman: Science **257**, 189 (1992);
   E. Betzig, J. Trautman, R. Wolfe: Appl. Phys. Lett. **61**, 142 (1992) 23
2. B.D. Terris, H.J. Marnin, G.S. Kino: Appl. Phys. Lett. **65**, 388 (1994) 23
3. J. Tominaga, T. Nakano, N. Atoda: Appl. Phys. Lett. **73**, 2078 (1998) 23, 28, 31
4. J. Tominaga, T. Nakano, N. Atoda: Proc. SPIE **3467**, 282 (1998) 23, 28, 31
5. J. Tominaga, H. Fuji, A. Sato, T. Nakano, T. Fukaya, N. Atoda: Jpn. J. Appl. Phys. **37**, L1323 (1998) 23, 28, 31
6. T. Nakano, A. Sato, H. Fuji, J. Tominaga, N. Atoda: Appl. Phys. Lett. **75**, 151 (1999) 23, 28, 31
7. J. Tominaga, H. Fuji, A. Sato, T. Nakano, N. Atoda: Jpn. J. Appl. Phys. **39**, 957–961 (2000) 24, 31

8. J. Tominaga, D. Büchel, T. Nakano, T. Fukaya, N. Atoda, H. Fuji: Proc. SPIE **4081**, 86–94 (2000) 24, 31
9. T. Fukaya, J. Tominaga, T. Nakano, N. Atoda: Appl. Phys. Lett. **75**, 3114 (1999) 24, 31
10. F. H. Ho, W. Y. Lin, H. H. Chang, Y. H. Lin, W. C. Liu, D. P. Tsai: Jpn. J. Appl. Phys. **40**, 4101 (2001) 24
11. D.P. Tsai, W. C. Lin: Appl. Phys. Lett. **77**, 1413 (2000) 24
12. D.P. Tsai, C.W. Yang, W.C. Lin, F.H. Ho, H.J. Huang, M.Y. Chen, T.F. Tseng, C.H. Lee, C.J. Yeh: Jpn. J. Appl. Phys. **39**, 982 (2000) 24
13. W. C. Liu, C. Y. Wen, K. H. Chen, W. C. Lin, D. P. Tsai: Appl. Phys. Lett. **78**, 685 (2001) 24, 31
14. T. Fukaya, D. Büchel, S. Shinbori, J. Tominaga, N. Atoda, D. P. Tsai, W. C. Lin: J. Appl. Phys. **89**, 6139 (2001) 24
15. D.P. Tsai, Y.Y. Lu: Appl. Phys. Lett. **73**, 2724 (1998) 24
16. D.P. Tsai, C. W. Yang, H. E. Jackson: 5th Int. Conf. on Near-Field Optics and Related Techniques, Shirahama, Japan, Dec. 6–10, 1998, Tech. Dig., p. 474 24
17. The Nanoscope IIIa from Digital Instruments Inc., Santa Barbara, CA 93117, USA 24
18. D.P. Tsai, C.W. Yang, S.Z. Lo, H.E. Jackson: Appl. Phys. Lett. **75**, 1039 (1999) 24, 25
19. N.H. Lu, W.C. Lin, D.P. Tsai: J. Microsc. **202**, 172 (2001) 25
20. H. Raether: *Surface Plasmons on Smooth and Rough Surfaces and on Gratings* (Springer, Berlin, Heidelberg 1988) 27
21. S. Kawata (Ed.): *Near-Field Optics and Surface Plasmon Polaritons*, Topics Appl. Phys. **81** (Springer, Berlin, Heidelberg 2001) 27
22. V.M. Agranovich, D.L. Mills: *Surface Polaritons* (North-Holland, Amsterdam 1982) 27
23. D.P. Tsai, J. Kovacs, Z. Wang, M. Moskovits, J.S. Suh, R. Botet, V.M. Shalaev: Phys. Rev. Lett. **72**, 4149 (1994) 31
24. V.M. Shalaev (Ed.): *Optical Properties of Nanostructured Media*, Topics Appl. Phys. **82** (Springer, Berlin, Heidelberg 2002) 31

Super-RENS Media Using Alternative Recording Systems

Jooho Kim[1,2]

[1] Laboratory for Advanced Optical Technology (LAOTECH), National Institue of Advanced Industrial Science and Technology (AIST), Tsukuba, 305-8562, Japan
[2] Digital Media R&D Center, Samsung Electronics Co., Ltd. Suwon, 442-742, Korea
j.h.kim@aist.go.jp

Abstract. The super-RENS system has been studied with phase change recording materials such as GeSbTe and AgInSbTe, and surface-plasmon near-field signals have been analyzed by the reflectivity difference caused by the phase change. In this paper, we introduce alternative recording systems in the super-RENS. At first, it was investigated whether magneto-optical signals also keep their Kerr polarization feature in the surface plasmon near-field, and can be applied to the super-RENS system. As another topic, the reactive diffusion recording mechanism and its super-RENS application is described in order to improve the resolution limit and thermal stability of the media.

1 Super-RENS with a Magneto-Optical Recording System

Using a silver oxide thin film of a nonmagnetic mask layer, the magneto-optical (MO) recording signals were enhanced by surface plasmons. A resolution of less than 170 nm was achieved by the near-field coupling between a light-scattering center generated in the AgO_x film and the light polarization of MO marks. The electrical field intensity was confirmed by a finite-differential time-domain (FDTD) computer-simulation. The optimization of the disk structure and materials, and the effects of readout power were also investigated.

1.1 Introduction

Near-field optical data storage is an attractive research field for achieving high-density recording beyond the optical diffraction limit. Methods using a solid immersion lens (SIL) and a scanning near-field optical microscope (SNOM) have been studied for retrieving magneto-optical (MO) signals by near-field coupling [1,2,3,4,5,6,7,8]. On the other hand, a super-resolution near-field structure (super-RENS) has now been developed in phase change (PC) recording [9,10]. It is thought that a SIL together with the super-RENS

J. Tominaga and D. P. Tsai (Eds.): Optical Nanotechnologies,
Topics Appl. Phys. **88**, 35–49 (2003)

could be a more promising technique at present for near-field applications compared to other techniques. In MO disks, magnetic super-resolution (MSR) and magnetic amplifying magneto-optical system (MAMMOS) techniques have been proposed with special readout and recording magnetic layers consisting of GdFeCo and DyFeCo [11,12]. However, research using nonmagnetic mask layers to enhance magnetic signal intensities has not been carried out for near-field recording. As reported recently [13], a transparent aperture (Sb)-type super-RENS MO disk hardly enhances the signal intensity but the light-scatter-center-type (AgO_x) super-RENS MO disk, on the other hand, overcomes the theoretical resolution limit and can achieve readout signals up to 150 nm mark length. A carrier-to-noise ratio (CNR) at a mark length of 300 nm, which is below the resolution limit, is more than 20 dB, compared to 0 dB in conventional disks. This means that the Ag particles or their clusters generated in AgOx film can enhance MO signals by approximately 100 times of those of the conventional MO disks, and a nonmagnetic light-scattering-center can conserve the polarized MO signals. In this work, we attempted to optimize the structure and the material used for the near-field super-RENS MO disk.

1.2 Preparation

To identify the optimized recording conditions, disk structure and materials, we prepared several super-RENS MO disks using a silver oxide (AgOx) mask layer. This optical mask layer is nonmagnetic, but has optically nonlinear characteristics. For the experiments, we prepared sample disks with different intermediate layer thicknesses from 10 to 200 nm and using different dielectric layer species such as SiN_x of 60–100 nm thickness. The basic structure of a typical AgOx super-RENS disk in this work is depicted in Fig. 1. All films were deposited by RF magnetron sputtering with composite targets except for the AgOx film. The AgOx film was produced by RF-reactive magnetron sputtering using a pure Ag target and a gas mixture of O_2 and Ar. In the case of the AgOx film, the ratio of the O_2 gas flow to the total gas flow was about 0.5, and the k value of the refractive index was about 1.5. To evaluate the recording and readout characteristics of the disks, a Nakamichi OMS-2000 MO disk-drive tester was used which had a laser wavelength of 680 or 780 nm and a lens numerical aperture of 0.55 or 0.53, whose resolution limit is about 310 or 370 nm, respectively. The measuring conditions of the sample disks were a linear velocity of 6.0 m/s, a magnetic intensity of 200 Oe, and a bandwidth of 30 kHz. The characteristics of the readout signal intensities were measured using different laser wavelengths, intermediate layer thicknesses and dielectric layer species.

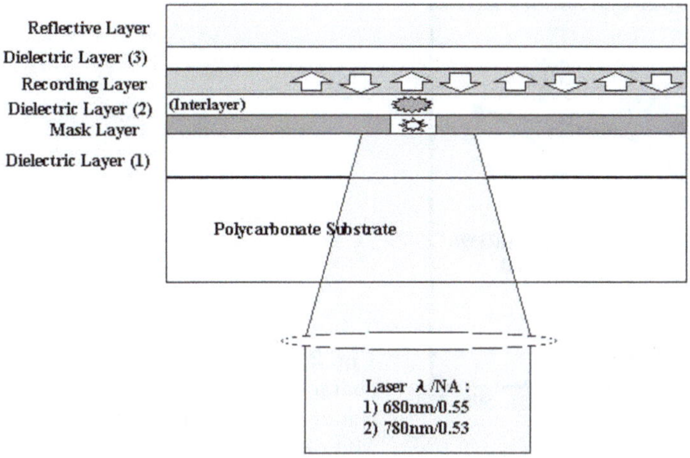

Fig. 1. Sample MO-structure and film thickness for light-intensity modulation (LIM) recording

1.3 Results and Discussion

To obtain the super-RENS effect in MO disks, we determined the characteristics of the super-RENS MO disk with a AgO_x mask film at different readout powers. As shown in Fig. 2, signal enhancement was obtained up to 170 nm mark length at a high readout power of 3.0 mW. In contrast, the same disk did not show any signal enhancement at a low readout power of 1.0 mW. This suggests that the AgOx layer changes partially to heat-generated Ag by the strong laser heat, as reported previously [14]. Although the MAMMOS technique can amplify crescent-shaped domains of 100 nm breadth, MO signals from circular magnetic domains with a size of 300 nm cannot be enhanced by applying MAMMOS [15]. However, it was determined in this work that the circular domains can be detected by near-field coupling up to less than 170 nm using an optically nonlinear AgOx mask layer. The circular shape could be confirmed using a magnetic-force microscope (MFM). In MAMMOS, the length of the crescent-shaped magnetic domains in the radial direction is usually about 5 times more than that of the breadth. It is difficult to increase the radial recording density. On the other hand, the marks recorded in this work are circular and the area density is much larger than that in MAMMOS. To examine the wavelength effect, we measured the signal intensities of one sample disk using two different disk drives with a wavelength of 680 nm (NA 0.55) or 780 nm (NA 0.53). As shown in Fig. 3, in the case of a 680 nm wavelength, we could readout up to 170 nm mark length. In the case of a 780 nm wavelength, on the other hand, the resolution was up to 200 nm mark length. On the basis of this result, a shorter wavelength retrieves much smaller marks. We also recognized that the conservation ability of polarized MO signals in the AgOx light-scattering type is much higher than that of the

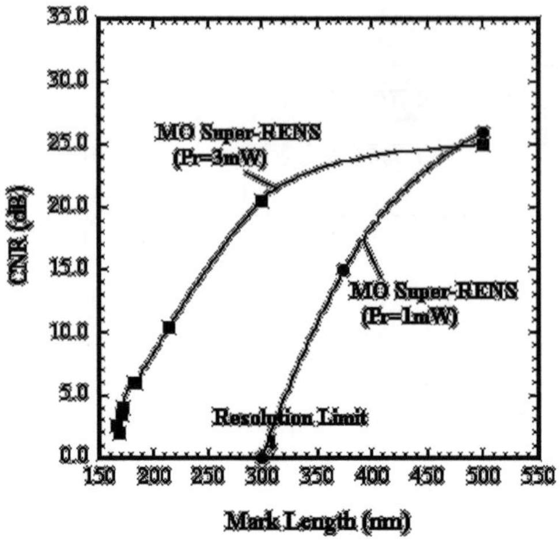

Fig. 2. CNR and mark length characteristics of light-scattering center (LSC)-type super-RENS MO disk

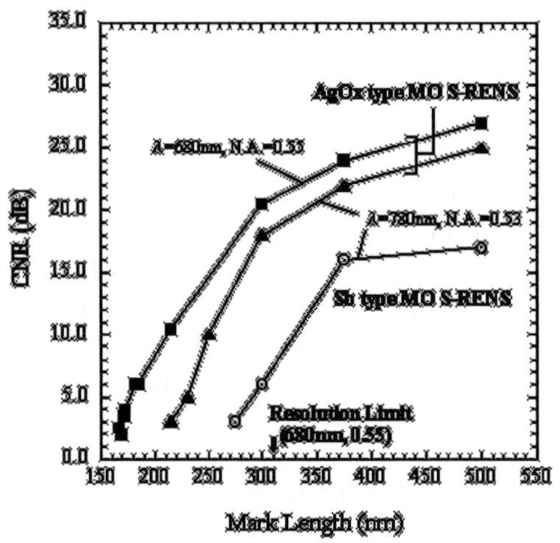

Fig. 3. Effect of optical system (wavelength and NA) and mask-layer type

Sb transparent aperture type super-RENS MO disks from comparison of the CNR.

To reveal signal enhancement by the light-scattering center in super-RENS disks, the electric field intensity was estimated using Sb and AgOx as the mask-layer by computer simulation with the finite-differential time-domain (FDTD) method. The light-scattering center and transparent aperture were assumed as a Ag particle and as an Sb amorphous aperture. The size of the aperture and scattering center were fixed at 200 nm in this simulation. As shown in Fig. 4, the z-component of the electric field of the light-scattering center is about 20 times stronger than that of the transparent aperture underneath the mask layers. This result agrees well with the experimental results, although the polarization components of the MO layer were not considered in this simulation. Since we developed the super-RENS disk using AgOx, we

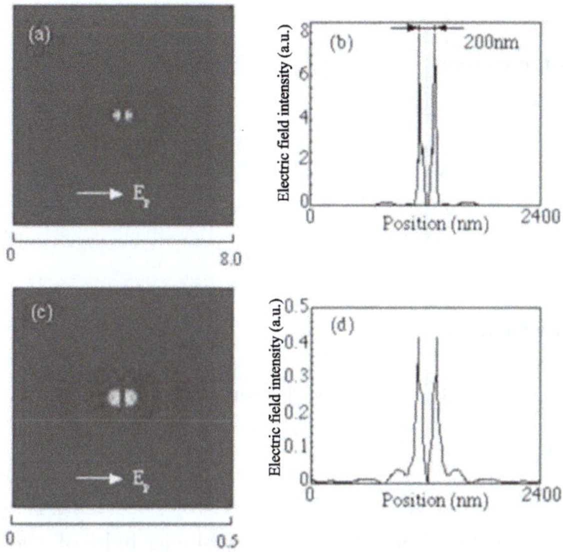

Fig. 4. Electric field of z-components beneath mask layers calculated by FDTD. (**a**) Near-field distribution scattered in the x–y plane of super-RENS using Sb film and (**b**) field intensity. (**c**) Near-field distribution scattered in the x–y plane of super-RENS using Sb film and (**d**) field intensity. E_{p} is the direction of the incident light polarization

have looked for evidence of surface plasmon or local plasmon enhancement. As is well known, surface plasmons only appear under certain conditions at the interface between a metal and a dielectric material. In order to generate the surface plasmons effectively we have to use a total reflection method or use gratings or rough surfaces in general [16]. *Shalaev* has shown that aggregated Ag particles or their clusters with a size range of $\sim$ 100 nm may also generate plasmons, resulting in huge local-field enhancements [17,18]. The strong electrical field enhancement, in some cases, can be up to 10^{10}, which is four orders of magnitude larger than the average for the unique film. The enhancement effect was actually observed as surface-enhanced Raman images by NSOM. On the other hand, according to the refractive-index evaluation of AgOx in our recent work, the refractive index was shifted from its initial AgOx (mostly Ag_2O) state with a value of $2.5 + 1.5$j to a more Ag-rich state with a value of $0.7 + 3.7$j by increasing the readout power, generating a light-scattering center. The Ag-rich local phase may be compatible with the localized states of Ag mentioned by Shalaev. Moreover, as the mark distance recorded on a recording film (MO or PC) gets closer to the near-field region, a portion of the incident light may be in the mark train. This effect occurs along the train because the electric field is prohibited from transmitting the film by diffraction theory, but the field perpendicular to the train can transmit the film. The absorbed or trapped field is thus transferred into a local

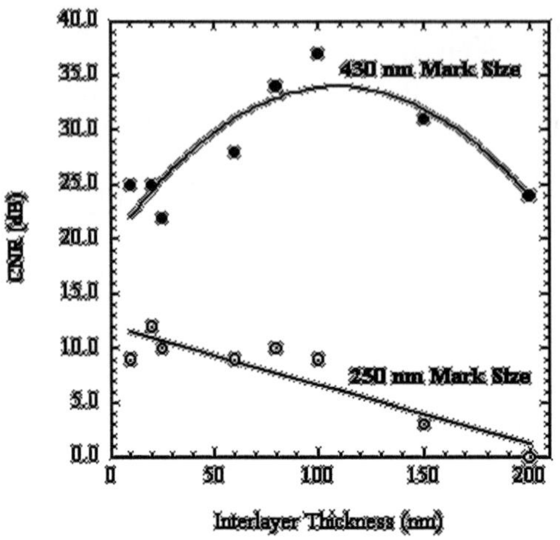

Fig. 5. Effect of interlayer thickness between mask and recording layers

plasmon, which is scattered to the far field by the strong local field of the aggregated Ag particles. It should be noted that the electric field reflected or transmitted from super-RENS disks is partly polarized. This effect plays an important role for signal deviation in the super-RENS MO, although the details are not clear at the moment. Figure 5 shows the effect of the interlayer thickness, which is the layer thickness between the mask and the recording layer. In the case of a 430 nm mark-length signal in the far field, the maximum appears at approximately 100 nm thickness. This is caused by the reflectivity change due to the interference of the multilayer structure. The signal characteristic of a 250 nm mark length in the near field decreases almost linearly, which means an exponential decrease of the near-field signal.

We confirmed that the MO signal could be retrieved by the near-field super-RENS effect with a light-scattering center (AgO_x) and it was found that the nonmagnetic layer could conserve the Kerr polarization. The results indicate the possibility of enhancing the resolution of MO circular signals further up to less than 100 nm mark length by optimizing the structure and the materials.

2 Super-RENS with Reactive Diffusion Recording System

Reactive recording was achieved with typical rare-earth transition metals (RE-TM) for magneto-optical (MO) recording. Almost the same carrier-to-noise ratio (CNR) and much higher modulation were obtained by the reactive recording, compared with conventional phase change (PC) recording. By applying this recording material to a super-resolution near-field structure for terabyte recording, the CNR below 100 nm mark-length signal, readout durability and power margin were greatly improved. To identify the record-

Table 1. Sample structure and film thickness

Layers	Structure I (nm)	Structure II disk, (nm)	Structure III disk, (nm)
Substrate	glass	polycarbonate	polycarbonate
Dielectric 1	$ZnS–SiO_2/Si_3N_4$ (100)	$ZnS–SiO_2/Si_3N_4$ (170)	$ZnS–SiO_2$ (85)
Mask	–	–	AgO_x (15)
Dielectric 2	–	–	$ZnS–SiO_2$ (70)
Recording	TbFeCo/GeSbTe (20)	TbFeCo/GeSbTe (20)	TbFeCo/GeSbTe (20)
Dielectric 3	$ZnS–SiO_2/Si_3N_4$ (50)	$ZnS–SiO_2/Si_3N_4$ (65)	$ZnS–SiO_2$ (65)
Reflective	–	–	Ag (50)

ing mechanism, we examined the magnetic and thermo-optical properties, finding that the film properties of amorphous RE-TM are sharply changed at approximately 773 K by crystallization and thermal-activated reaction with dielectric layers.

2.1 Introduction

The recording mechanism of magneto-optical (MO) data storage is magnetization reversal by an external magnetic field and laser heating (thermo-magnetic recording), which results in magnetic Kerr and Faraday rotation of the incident light [19]. In contrast, phase change (PC) recording uses the transformation from the amorphous to the crystalline phase by external laser heating, which results in a reflectivity change [20]. In this work, by using reactive recording with a rare-earth transition metal (RE-TM), which is a typical MO recording material, we tried to record and retrieve signals in both MO and PC drives with the same disk. Recently, to achieve terabyte-capacity data storage, a super-resolution near-field structure (super-RENS) has been proposed and succeeded in retrieving signals with less than 100 nm mark length using a AgO_x (light-scattering center type) or Sb (transparent-aperture type) mask layer [21,22,23]. However, the super-RENS disk needs further improvements; higher CNR below 100 nm mark-length signal, better durability of readout cycles and larger readout power margin. To improve the above characteristics, a double-mask layer structure, blue-laser drive system, and silver (Ag) or oxygen (O_2) doped GeSbTe materials have been introduced [24,25,26]. But sufficiently good results have not been obtained yet. In the following sections we describe how we carried out reactive recording by applying RE-TM for making the above improvements and report on the reaction process and mechanism. To elucidate the mechanism, we also examined the optical constants and the magnetic and thermo-optical properties.

2.2 Preparation

We prepared samples deposited by rf magnetron sputtering with composite targets except for the AgO_x film, which was produced by rf reactive mag-

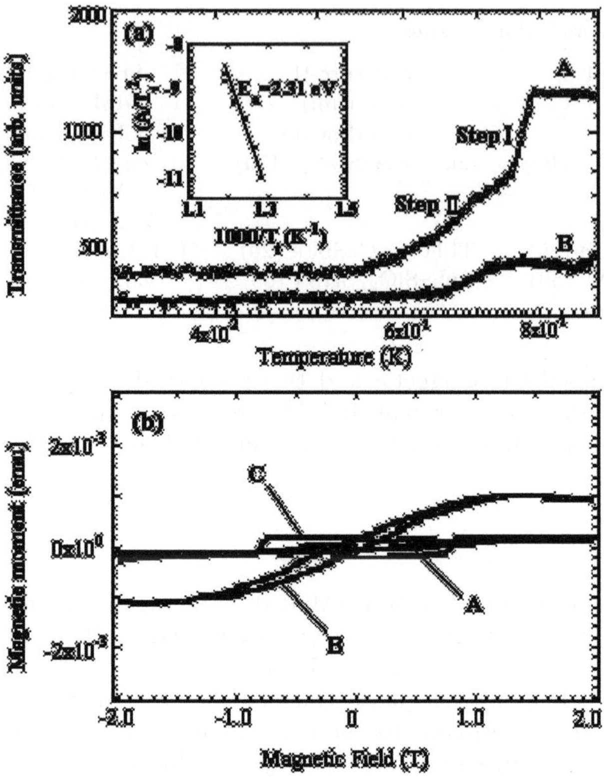

Fig. 6. (**a**) Transmittance change of the samples as a function of temperature (structure I): A, $ZnS–SiO_2$/MO/$ZnS–SiO_2$; B, Si_3N_4/MO/Si_3N_4. Inset: Kissinger's plot for activation energy of transition. (**b**) Magnetic hysteresis curves of as-deposited (C) and 873 K-heated (A, B) samples

netron sputtering with a pure Ag target and a gas mixture of O_2 and Ar on the polycarbonate substrate (Table 1). To examine the recording and readout characteristics, a Pulstec DDU-1000 PC drive tester (laser wavelength, $\lambda = 635$ nm and lens numerical aperture, NA = 0.6) and a Nakamichi OMS-2000 MO drive tester ($\lambda = 780$ nm, NA = 0.53) were used. Sample disks were rotated at a constant linear velocity of 3.0 m/s. The CNRs of the readout signals were measured with a bandwidth of 30 kHz. The magnetic intensity for recording and erasing the MO signals was 200 Oe. Magnetic and thermo-optical properties and the optical constants were estimated with a vibrating sample magnetometer (VSM, Oxford Instrument, MagLab-VSM-12 V), an ellipsometer (Mizojiri Optical Co., DHA-OLX/S4M), a multi-channel photodetector (Hamamatsu Photonics, PMA-11) and a spectrophotometer (Shimadzu, UV-2500 PC).

2.3 Results and Discussion

We designed the structure by considering the crystallization of the RE-TM and the reaction between the RE-TM and the dielectric layers (Table 1). To reveal the reaction process and mechanism, we investigated the thermo-optical property of the samples having a three-layer-structure I (Fig. 6a). In Fig. 6a, curve A represents the transmittance change versus the temperature increase of the sample sandwiched between the ZnS–SiO_2 dielectric layers. It was found that there was no change up to around 573 K. Following a relatively slow increase from 573 K to 763 K (Step II), a steep increase of transmittance occurred from 763 K to 783 K (Step I). The transmittance of the as-deposited and 783 K-heated samples greatly changed from 400 to 1300 arbitrary units; the difference was about 900. This behavior is similar to a completely oxidized RE-TM single-layer sample. In the case of the PC sample, the difference was 150. Comparing the as-deposited sample to the 873 K-heated sample, the optical constants of the RE-TM film at 633- nm wavelength, n, k (refractive index: n, extinction coefficient: k), were changed from 3.3, 4.0 (as-deposited) to 2.2, 0.3 (873 K-heated), respectively. This means that the sample becomes almost transparent. It is thought that the above change is related to the crystallization and reaction of the RE-TM. It was reported that RE-TM is first crystallized to α-TM and α-RE and is then followed by the crystallization of the intermetallic compound of RE-TM in the amorphous RE-TM film [27]. It is supposed that the reaction includes both the sulfuration and oxidation of RE-TM (mainly sulfuration). It has been reported that sulfur in ZnS–SiO_2 is easily released and diffuses into a PC (GeSbTe) recording layer under external heating; this significantly changes the optical constants and readout cyclability [28,29]. The oxidation of amorphous RE-TM has also been mentioned by several researchers: the preferential oxidation of Tb atoms (strong affinity for oxygen) and the oxidation reaction with the SiO_2 dielectric layers of RE-TM [30,31,32]. It is thought that the crystallization of the amorphous RE-TM film occurs dominantly during the range of the first step (573 K–763 K) from the observation of a slow increase of the transmittance. In the region of the second step (763 K–783 K), the thermal-activated reaction of the film seems to occur dominantly. We also tried to obtain the activation energy for the transition from the Kissinger formula $\ln(A/\mathrm{T}_\mathrm{t}^2)=E_\mathrm{a}/K_\mathrm{b}T_\mathrm{t} + C)$, where A is the heating rate, T_t is the transition temperature at heating rate A and KB is the Boltzmann constant (inset of Fig. 6a) [33]. The activation energy for the transition of the sample was about 2.31 eV, which was similar to the value of 2.51 eV for GeSbTe phase-change material [34]. As shown in Fig. 6a, curve B, we could not find a large increase of transmittance in a RE-TM film sandwiched with Si_3N_4 dielectric layers with increasing temperature. This means that free sulfur and non-bridging oxygen (NBO) within rf-sputtered ZnS–SiO_2 dielectric layers plays an important role in the reactive recording of a RE-TM film [35]. Examining the photograph taken by transmission electron microscope (TEM), it was found that the RE-TM thin film was crystallized

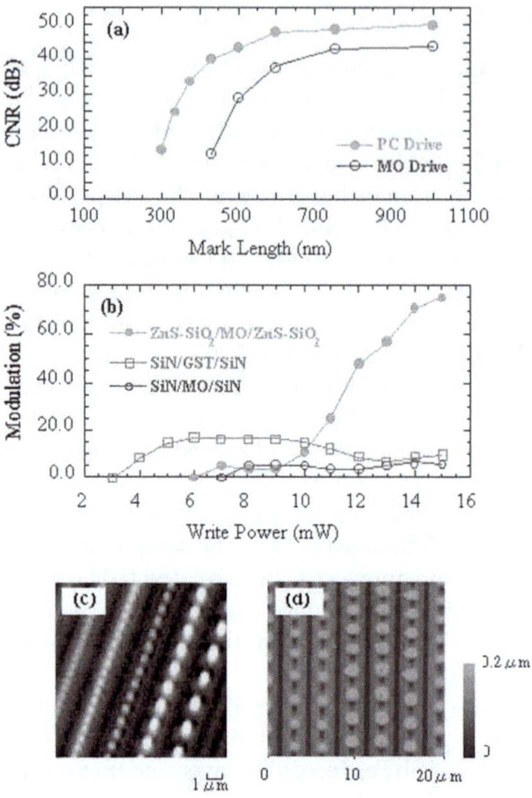

Fig. 7. Conventional disk characteristics: (**a**) frequency response of the sample disk (structure III) in MO and PC drives; (**b**) modulation of different sample disks with the three-layer structure II; (**c**) optical microscope photograph (transmission mode) of different mark-length signals; (**d**) AFM image of the modulation-measure sample

and reacted, the layer diffused into the ZnS–SiO_2 dielectric layers, and volume expansion occurred in the 783 K-heated sample. We also investigated the magnetic properties of the samples to confirm the reacted-state of the RE-TM film. The magnetic properties (coercivity: H_c, magnetization: B_s) of the heated sample (873 K) sandwiched with ZnS–SiO_2 dielectric layers was the same as that of completely oxidized RE-TM film, as shown in the hysteresis curve of Fig. 6b, curve A. The heated sample (873 K) sandwiched with Si_3N_4 dielectric layers, however, kept its magnetic properties (lower H_c and larger B_s than those of the as-deposited sample; see curve C) as shown in the hysteresis curve of Fig. 6b, curve B.

Next, we applied this reactive material to a data storage disk structure. The frequency response in the PC and MO drives of the sample disk is shown in Fig. 7a. The structure consisted of the recording, dielectric, mask and reflective layers (structure III) with the same thickness, except for the polycarbonate substrate thickness (0.6 mm for PC drive, 1.2 mm for MO drive). The disks showed almost the same CNR versus mark-length characteristics, considering the drive characteristics, in which the resolution limits ($\lambda/4 \times NA$) are 270 nm and 370 nm for the PC and MO drive, respectively. The readout laser power was 1.0–1.5 mW and the recording laser power was 5.0–6.0 mW. This means that the disk for the PC and MO drives can be identified with the same disk having the above structure. Figure 7b represents the modulation characteristics, which means the signal reflection difference (voltages) between the recorded and unrecorded marks in the PC drive tester. The structure consists of RE-TM (MO, TbFeCo) or PC (GeSbTe) recording layers sandwiched by dielectric layers: ZnS–SiO_2 or Si_3N_4 (three-layer, structure II). As shown in Fig. 7b, the RE-TM layer sandwiched by ZnS–SiO_2 had a very large modulation (more than 75 %) in comparison with that of three-layer-PC-structure (approximately 20 %). The RE-TM layer sandwiched by Si_3N_4, on the other hand, had a small modulation of less than 10 %. According to these results, it is confirmed that the reaction between the RE-TM and ZnS–SiO_2 layers produces the large contrast marks in reflection. Figure 7c shows the marks recorded by the reactive recording with a transmission-mode optical microscope. It is well recognized that the recorded marks are transparent. Examining the surface with an atomic force microscope (AFM), it was also found that the recorded marks have a convex shape, which supports a volume expansion by the reaction (Fig. 7d).

We applied the RE-TM recording layer to a super-resolution near-field structure (super-RENS) disk and compared it with a PC super-RENS disk with a AgO_x mask layer. As shown in Fig. 8a, the CNR (24 dB) of the sample disk by RE-TM (MO) reactive recording was twice as high as that (12 dB) of the PC super-RENS disk for 100 nm mark-length signals. Examining the spectrophotometer results, it was found that the samples of three-layer-structure I have a similar large reflectivity change (20 %) in the blue-laser region (405 nm), compared with 23 % in the red-laser region (635 nm) between as-deposited and 873 K-heated samples. Therefore, we believe that we can apply the sample disks to the blue-laser drive system as well as the red-drive system and should expect to have better CNRs in both cases. Figure 8b shows the readout durability of the sample disk and PC super-RENS disk. The crystallization temperature of the PC material (GeSbTe) is around 453 K [36], and the temperature is close to the AgO_x decomposition temperature (around 433 K), which corresponds to the super-RENS readout power (approximately 3.0 mW) [37,38]. In the high-power readout process, the non-recorded area is gradually degraded from the amorphous (non-recorded) to the crystalline (recorded) state, which results in the CNR drop. Therefore, the super-RENS disk needs recording materials with a much higher transition temperature. As the reaction temperature of the RE-TM recording layer is above 573 K (Fig. 6a), the sample disk indicates excellent readout durability. The readout power margin is also improved (Fig. 8c). In

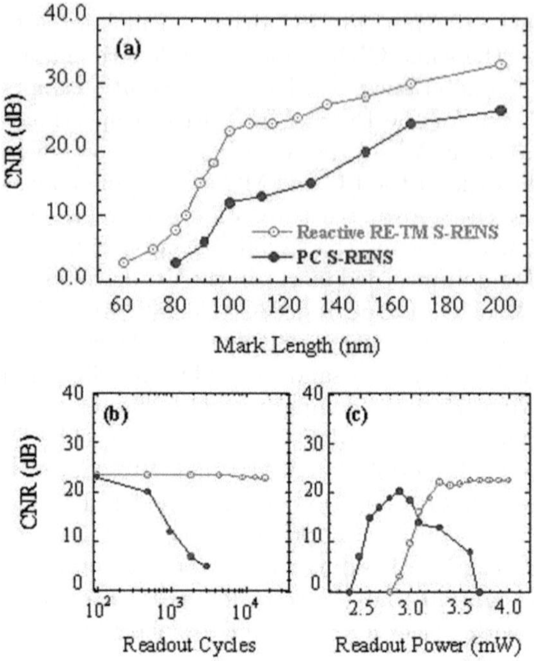

Fig. 8. Super-resolution near-field structure disk characteristics: (**a**) frequency response; (**b**) readout durability; (**c**) readout power margin for the sample disk (150 nm signals) and PC super-RENS disk (200 nm signals)

the case of a PC super-RENS disk, by increasing the power beyond the AgO_x decomposition temperature, the transition to the crystalline phase is accelerated; the CNR is, in turn, sharply reduced. The RE-TM reactive sample disk, however, provides a better power margin in the region higher than 3.3 mW. It is thought that this also depends on the high transition temperature of the reactive recording material: RE-TM.

Reactive recording with RE-TM was successfully achieved and applied to optical data storage. It was found that the CNR, readout durability and power margin were greatly improved by the reactive recording method. It is supposed that the recording mechanism is due to the crystallization and thermal-activated reaction of RE-TM to generate the large change of transmittance and volume. However, the details are not clear at the moment and we need further work to elucidate the process. The results in this work may indicate the potential of reactive recording between recording and dielectric layers. We expect to find other materials that react with dielectric layers, to enable other recording mechanisms.

References

1. J. Qui, K. Miura, K. Hirao: Jpn. J. Appl. Phys. **37**, 2263 (1998) 35
2. T. Ohta, K. Nishiuchi, K. Narumi, Y. Kitaoka, H. Ishibashi, N. Yamada, T. Kozaki: Jpn. J. Appl. Phys. **39**, 770 (2000) 35
3. A. Partovi, D. Peale, M. Wutting, C. Murry, G. Zydzik, L. Hopkins, K. Baldwin, W. Hobson, J. Wynn, J. Lopata, L. Dhar, R. Chichester, J. Yeh: Appl. Phys. Lett. **75**, 1515 (1999) 35
4. K. Hirota, T. Milster, Y. Zhang, J. Erwin: Jpn. J. Appl. Phys. **39**, 973 (2000) 35
5. W. Hung, M. Mansuripur: Appl. Opt. **39**, 302 (2000) 35
6. F. Guo, T. Schlesinger, D. Stancil: Appl. Opt. **39**, 324 (2000) 35
7. S. Jiang, H. Ohsawa, K. Yamada, T. Pagaribuan, M. Ohtsu, K. Imai, A. Ikai: Jpn. J. Appl. Phys. **31**, 2282 (1992) 35
8. M. Lee, M. Kourogi, T. Yatsui, K. Tsutsui, N. Atoda, M. Ohtsu: Appl. Opt. **38**, 3566 (1999) 35
9. J. Tominaga, H. Fuji, A. Sato, T. Nakano, N. Atoda: Jpn. J. Appl. Phys. **39**, 957 (2000) 35
10. H. Fuji, J. Tominaga, L. Men, T. Nakano, H. Katayama, N. Atoda: Jpn. J. Appl. Phys. **39**, 980 (2000) 35
11. K. Itoh, H. Yoshimura, K. Ogawa: Jpn. J. Appl. Phys. **39**, 714 (2000) 36
12. H. Awano, M. Sekine, M. Tani, N. Kasajima, N. Ohta, K. Mitani, N. Takagi, S. Sumi: Jpn. J. Appl. Phys. **39**, 725 (2000) 36
13. J.H. Kim, D. Büchel, T. Nakano, J. Tominaga, N. Atoda, H. Fuji, Y. Yamagawa: Appl. Phys. Lett. **77**, 1774 (2000) 36
14. J. Tominaga, D. Büchel, T. Nakano, H. Fuji, T. Fukaya, N. Atoda: Proc. SPIE **4081**, 86 (2000) 37
15. H. Awano, H. Shirai, N. Ohta, A. Yamaguchi, S. Sumi, K. Torazawa: Nippon Ouyoh Jiki Gakkaishi **22**, 337 (1998)[in Japanese] 37
16. H. Raether: *Surface Plasmons* (Springer, Berlin, Heidelberg 1988) 39
17. V.M. Shalaev: *Nonlinear Optics of Random Media*, Springer Tracts Mod. Phys. **158** (Springer, Berlin, Heidelberg 2000) p. 141 39
18. V.M. Shalaev (Ed.): *Optical Properties of Nanostructured Random Media*, Topics Appl. Phys. **82** (Springer, Berlin, Heidelberg 2002) p. 52 39
19. M. Mansuripur: *The Physical Principles of Magneto-optical Recording* (Cambridge University Press, Cambridge 1995) p. 638 41
20. M. Okuda, H. Naito, T. Matsushida: Jpn. J. Appl. Phys., **31**, 466 (1992) 41
21. J. Tominaga, T. Nakano, N. Atoda: Appl. Phys. Lett. **73**, 2078 (1998) 41
22. H. Fuji, J. Tominaga, L. Men, T. Nakano, H. Katayama, N. Atoda: Jpn. J. Appl. Phys. **39**, 980 (2000) 41
23. J.H. Kim, D. Büchel, T. Nakano, J. Tominaga, N. Atoda, H. Fuji, Y. Yamakawa: Appl. Phys. Lett. **77**, 1774 (2000) 41
24. J. Tominaga, J.H. Kim, H. Fuji, D. Büchel, T. Kikukawa, L. Men, H. Fukuda, T. Nagano, T. Fukaya, N. Atoda, A. Sato, A. Tachibana, Y. Yamakawa: Jpn. J. Appl. Phys. **40**, 1831 (2001) 41
25. L. Men, J. Tominaga, H. Fuji, Q. Chen, N. Atoda: Proc. SPIE **4085**, 204 (2001) 41
26. L. Men, J. Tominaga, H. Fuji, N. Atoda: Jpn. J. Appl. Phys. **39**, 2639 (2000) 41

27. S.R. Lee, A.E. Miller: J. Appl. Phys. **55**, 3465 (1984) 43
28. X.S. Miao, T.C. Chong, L.P. Shi, P.K. Tan, J.M. Li, K.G. Lim: Jpn. J. Appl. Phys. **40**, 1581 (2001) 43
29. N. Yamada, M. Otoba, K. Kawahara, N. Miyagawa, H. Ohta: Jpn. J. Appl. Phys. **37**, 2104 (1998) 43
30. R.B. Van Dover, E.M. Gyorgy, R.P. Frankenthal, M. Hong, D.J. Siconolfi: J. Appl. Phys. **59**, 1291 (1986) 43
31. F. Luborsky, J.T. Furey, R.E. Skoda, B.C. Wagner: IEEE Trans. Magn. **22**, 937 (1986) 43
32. T. Anthony, J. Brug, S. Naberhuis, H. Birecki: J. Appl. Phys. **59**, 213 (1985) 43
33. H.E. Kissinger: Anal. Chem. **29**, 1702 (1959) 43
34. L. Men, J. Tominaga, H. Fuji, T. Kikukawa, N. Atoda: Jpn. J. Appl. Phys. **40**, 1629 (2001) 43
35. I. Yasui: *Optical Materials* (Dainippon, Tokyo 1991) p. 24 [in Japanese] 43
36. J. Tominaga, T. Nakano, N. Atoda: Jpn. J. Appl. Phys. **37**, 1852 (1998) 45
37. J. Tominaga, H. Fuji, A. Sato, T. Nakano, N. Atoda: Jpn. J. Appl. Phys. **39**, 957 (2000) 45
38. D. Büchel, J. Tominaga, N. Atoda: J. Magn. Soc. Jpn. **25**, 240 (2001) 45

Metal-Doped Silver Oxide Films as a Mask Layer for the Super-RENS Disk

Takayuki Shima*, Dorothea Büchel, Christophe Mihalcea, Jooho Kim, Nobufumi Atoda, and Junji Tominaga

Laboratory for Advanced Optical Technology (LAOTECH), National Institute of Advanced Industrial Science and Technology (AIST),
1-1-1 Higashi, Tsukuba, 305-8562, Japan
t-shima@aist.go.jp

Abstract. Various kinds of metal (Co, Pd, Pt and Au) were doped into Ag_2O and AgO sputtered films to study its effect on the thermal decomposition process. The oxygen composition ratio was evaluated by the X-ray fluorescence spectroscopy method after annealing up to 260 °C. The optical transmittance change was measured during heating of the film to 600 °C. Noble metal doping was found to modify the AgO decomposition process, and the oxygen content decreased gradually compared to the undoped case. Super-RENS disks with a metal-doped AgO mask were prepared, and the laser power necessary for super-resolutional readout was evaluated. It slightly shifted to the higher-power side when the noble metal was doped, and this agrees with the modification of the decomposition process.

1 Introduction

The use of a super-resolution near-field structure (super-RENS) in optical disks has made possible the readout of small marks beyond the optical diffraction limit [1]. Silver oxide (Ag–O) film, recently used as a mask layer, decomposes into Ag clusters and oxygen by irradiation with a focused laser beam. Localized surface plasmons generated in these clusters are considered to provide the super-resolutional readout by interaction with small marks that have been recorded on a phase change storage layer [2]. Reducing the laser power for the decomposition is quite important in super-RENS disks, in order not to influence the closely spaced recording-layer characteristics. In addition to super-RENS applications, Ag–O layers have recently become attractive due to effects concerning surface-enhanced Raman spectroscopy [3] and intrinsic intermittent fluorescence [4]. Activation of the Ag–O film through its decomposition is essential in all these applications. It is thus important to study carefully the decomposition process, and also to search for a method to control it.

* Japan Science and Technology Corporation, Domestic Research Fellow

J. Tominaga and D. P. Tsai (Eds.): Optical Nanotechnologies,
Topics Appl. Phys. **88**, 49–58 (2003)

2 Silver Oxide Film

In this chapter, the thermal decomposition process of undoped and metal-doped Ag–O films is studied by composition analysis and optical transmittance measurements. Various kinds of metals (Co, Pd, Pt and Au) were doped into Ag–O film to examine on modification of the process. Super-RENS disks with a metal-doped mask layer were prepared, and the effect of the layer on the super-resolutional readout was evaluated.

2.1 Film Preparation

Silver oxide (Ag–O) films were prepared by the rf-magnetron reactive sputtering method. A 4 N–Ag target with a diameter of 76 mm was used, and the oxygen gas flow ratio was varied in the range from 0 to 0.75. The sputtering power and pressure were fixed at 200 W and 0.5 Pa, respectively, and the total gas flow of Ar and oxygen was maintained at 10 sccm. Further details of the preparation can be found elsewhere [2,5]. Films were deposited on glass, fused silica and Si substrates at room temperature. The film thickness was defined by a surface-texture measuring system, and it was about 100 nm for the samples used in the optical and thermal decomposition studies.

The composition ratio and optical properties of Ag–O film depend strongly on the oxygen gas flow ratio used for the film preparation. Figure 1 shows the result of the composition ratio variation, which was estimated using an X-ray fluorescence (XRF) spectrometer. The oxygen content gradually increased as a higher flow ratio was used, and it became nearly 50 at.% at the flow ratio of 0.75. It should be noted that there is some content variation that is less than ≈ 3 at.%. Figure 2 shows the refractive indices at a wavelength of 630 nm, which was obtained by a spectroscopic ellipsometry method. Similar and more detailed results are presented in [2,5]. The extinction coefficient k decreased as the flow ratio was increased, and the film became almost transparent when the ratio reached 0.5. *Schmidt* et al. have previously shown that

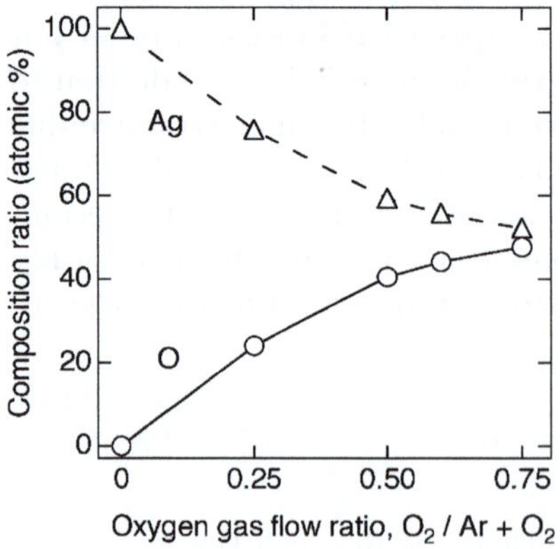

Fig. 1. Composition ratio of a Ag–O film as a function of the oxygen gas flow ratio

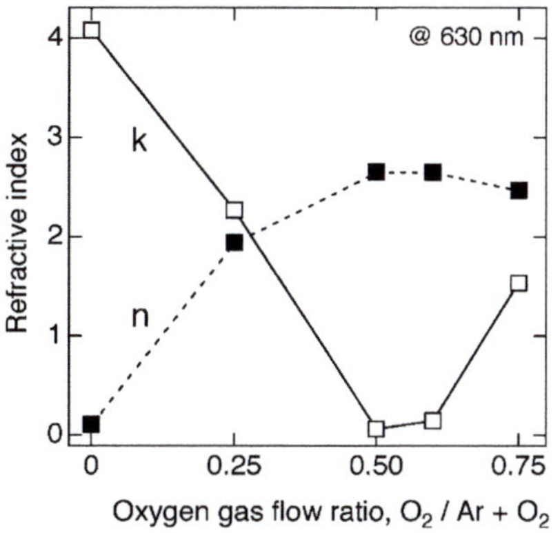

Fig. 2. Refractive indices of a Ag–O film as a function of the oxygen gas flow ratio

the refractive index of Ag_2O is $n = 2.5$ and $k = 0.11$ at 680 nm [6]. The transparent film obtained should thus reflect the formation of Ag_2O, though the XRF result indicated rather a higher oxygen content than expected. Further increase of the flow ratio to 0.75 raised the extinction coefficient k to about 1.5, and the film became opaque. *Büchel* et al. have previously identified from a Raman scattering spectroscopy study that such an opaque film contains AgO [5], and this is supported by the XRF result in Fig. 1.

2.2 Thermal Decomposition Process

The thermal decomposition process of Ag–O films was first studied by evaluating the composition ratio change after annealing. Figure 3 shows the result for both Ag_2O and AgO films prepared at the flow ratio of 0.5 and 0.75, respectively. Annealing was performed in air for 5 min using an oven that can heat up to 260 °C. For AgO film, the composition ratio changed drastically at around 130 °C, and became fairly similar to that of the Ag_2O film. This change derives from the thermal decomposition of AgO, and the film transformed to Ag_2O as an intermediate stage [5,7,8]. Above 130 °C, both Ag_2O and AgO films showed a slight decrease of the oxygen content, and this might indicate the partial decomposition of Ag_2O to Ag.

The process was also studied by measuring the transmitted light intensity change when heating the film sample. Figure 4 shows the result when the temperature was raised to 600 °C at a rate of 30 °C/min, and the transmittance was monitored at a wavelength of 635 nm. The AgO film showed an abrupt intensity increase at around 160 °C, and this corresponds to the transformation from opaque AgO to transparent Ag_2O. Above 160 °C, both the Ag_2O and AgO films showed a similar intensity variation, which decreased gradually until it turned to an increase at around 430 °C.

Figure 5 shows the optical transmittance spectra of the AgO film after heating at various temperatures. The sample was heated and cooled at a rate of 30 °C/min, and was kept for no time at the temperature reached. One

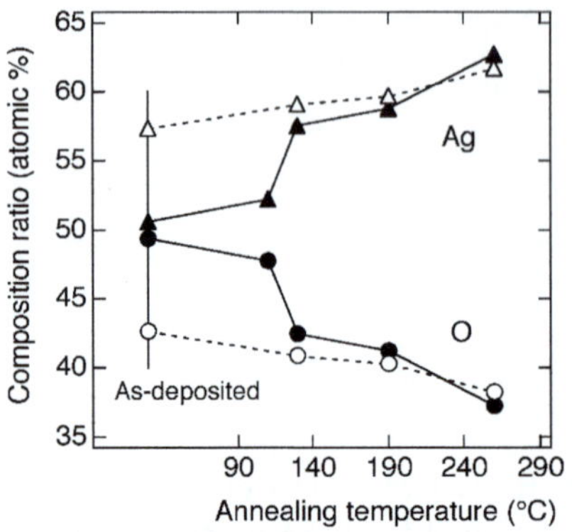

Fig. 3. Composition ratio of Ag–O films as a function of annealing temperature. *Open marks*: Ag_2O and *closed marks*: AgO

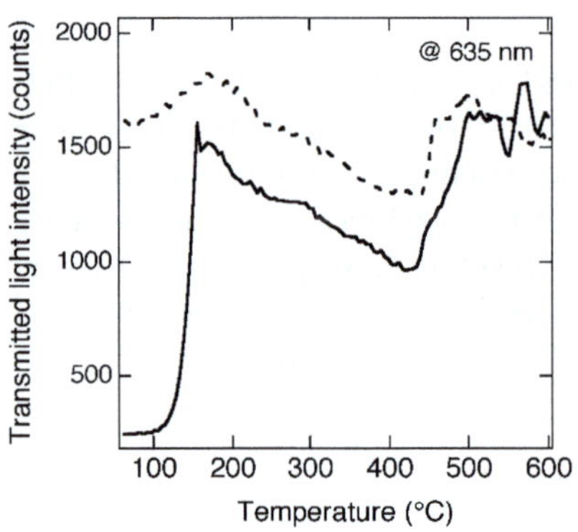

Fig. 4. Transmitted light intensity change of Ag–O films during heating. *Dashed line*: Ag_2O and *solid line*: AgO

can first confirm that the transmittance after heating at 160 °C in the visible wavelength region increased compared to as-deposited AgO, and became fairly close to that of as-deposited Ag_2O. A similar transmittance and spectrum shape can also be found after heating at 400 °C, and this suggests that the film still contained Ag_2O. Heating up to 600 °C made the spectrum shape different from that of Ag_2O or AgO, and the spectrum was less dependent on the wavelength. A small drop at 410 nm probably relates to the surface plasmon resonance absorption of silver particles [9], and this is evidence of Ag_2O decomposition to Ag.

A transmitted light intensity minimum at 430 °C in Fig. 4 is thus expected to relate closely to the Ag_2O decomposition process [7,8]. Particle formation can be recognized in the optical microscope image, and the spacing between the particles is probably the origin of the light intensity increase above 430 °C. Taking into account on the XRF result above 130 °C, the decrease of the light intensity found between 160 °C and 430 °C may correlate with the existence of Ag in Ag_2O. However, the evidence is still not clear at the moment, and a higher temperature was necessary to find a trace of Ag that corresponds to the total decomposition.

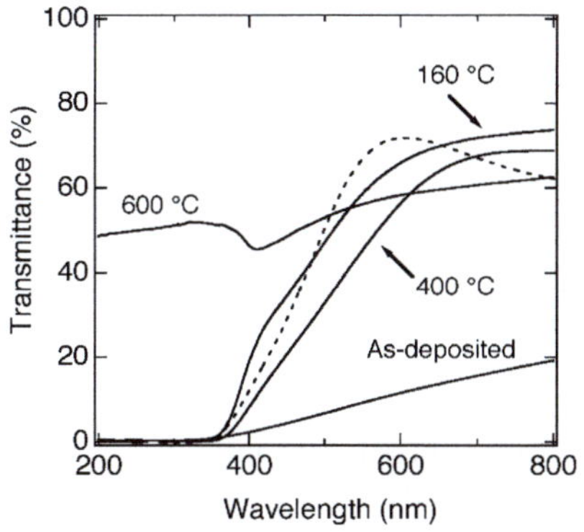

Fig. 5. Optical transmittance spectra of AgO film after heating at 160 °C, 400 °C, and 600 °C. *Dashed line*: as-deposited Ag_2O film

3 Metal-Doped Silver Oxide Film

Table 1 summarizes the kinds of metal doped into Ag–O film and its concentration estimated from XRF analysis. Doping was performed by attaching small metal chips (size: $5 \times 5\,mm^2$, purity: 3 N) on the Ag target, and they were simultaneously sputtered during the Ag–O film preparation.

3.1 Film Preparation

In this study, no precise control of the concentration was attemted, and the number of metal chips was fixed to two in all metal cases. The other preparation condition is identical to that for the undoped film in Sect. 2. The results showed that the doping was successfully made by this method with a concentration of the order of 0.1–1.0 at.%. In addition to the listed metals, Pd was also used as a doping metal; however, its concentration has not been well detected and identified at this moment.

3.2 Thermal Decomposition Process

The thermal decomposition process for metal-doped Ag–O film was first studied by measuring the transmitted light intensity change. This was because the method can easily be done and was found to trace well on the decomposition process of Ag–O film, as described in Sect. 2. The measurement conditions are identical to those used for the undoped films. Figure 6 shows the result

Table 1. Doped metals and their concentration in at.%

Doped metal	in Ag_2O	in AgO
Co	0.2	0.3
Pt	1.1	1.3
Au	1.6	1.8

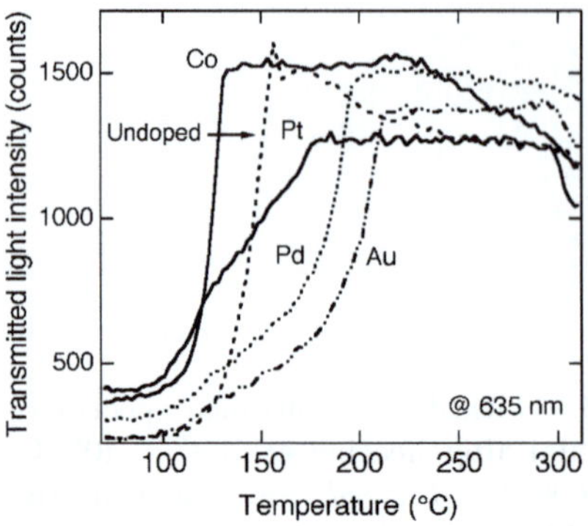

Fig. 6. Transmitted light intensity change of metal-doped AgO films during heating

for metal-doped AgO film in the temperature range of 60–310 °C. The curve labeled undoped is a duplicate of the one shown in Fig. 4.

The temperature at which the light intensity first reaches a maximum ($T_{\rm lm}$) was found to be dependent on the doping, which was about 160 °C for undoped film. When Pt, Pd and Au are doped, $T_{\rm lm}$ shifted higher by about 30 °C, 40 °C and 50 °C, respectively. On the contrary, $T_{\rm lm}$ shifted lower for Co doping by about 30 °C. The intensity curve shapes became like a plateau after the intensity reached its first maximum, and started to decrease at about 230 °C for Co and at about 300 °C for the noble metals. The process was also examined for the metal-doped Ag_2O film; however, no distinct difference in the intensity curve was recognized when compared to undoped Ag_2O in Fig. 4. For the AgO case and also the Ag_2O case, another intensity turning point at around 430 °C for the undoped film did not show any temperature shift due to the doping. This suggests that the method is not applicable at the moment for controlling the total decomposition temperature of Ag_2O to Ag.

To study in detail the modification of the transmitted intensity curves in Fig. 6, an XRF study was performed for the two doped cases in which $T_{\rm lm}$ shifted to lower and higher temperatures. The annealing conditions were identical to those described in Sect. 2. Figure 7 shows the oxygen composition ratio of Co-doped and Pt-doped AgO films when annealed. The result labeled undoped is a duplicate of the one shown in Fig. 3. For both doped cases, the ratio did not *drop* at around 130 °C and decreased gradually as the annealing temperature increased. This *continuous* decrease resulted in the plateau-like curve-shapes, since the Ag–O film is transparent over a certain oxygen content range [2,5]. This range includes at least 41–44 at.%, as described in Figs. 1 and 2. By Co doping, a decrease of the ratio was already noticeable at 110 °C, and this probably correlates with the lower temperature shift of $T_{\rm lm}$ in Fig. 6. By Pt doping, the film became more stable with respect to heating, and a higher temperature was necessary for oxygen removal. This shifted $T_{\rm lm}$ to the higher temperature side, and this is supposed to be also the case for other noble metals. The reason for the modification is still under investigation. A further study of Pt doping has recently been made, and

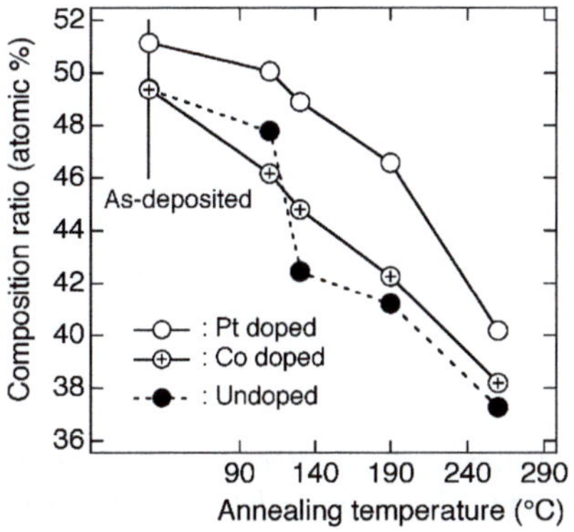

Fig. 7. Oxygen composition ratio of metal-doped AgO films as a function of annealing temperature

found that the oxygen ratio loss is strongly suppressed by increasing the Pt concentration to 2–5 at.% [10]. This is explained partly by that the decomposition property tends to be governed by it of Pt–O, which its decomposition is expected at higher temperature than Ag–O [11]. The Co-doping effect should be carefully studied to understand what starts the decomposition process at low temperature. However, the result in Fig. 7 also suggests that the modification itself was not remarkable, at least with the doping conditions used. It may still be necessary to examine whether there is any other doping material that is more efficient for achieving low-temperature decomposition.

4 As a Mask Layer in the Super-RENS Disk

Figure 8 shows the super-RENS disk structure prepared in this study. Details of its structure design and super-resolutional readout mechanism can be found in [1,2]. Both Ag_2O and AgO were used as a mask layer, and the metal doping was performed under the conditions described in Sect. 3. Since the refractive index of Ag–O film is slightly changed by the metal doping, the film thickness of the first dielectric layer was designed to keep the reflectivity of the disk nearly constant at a specific laser wavelength. Recording and retrieving of the mark signals was carried out using a disk-drive tester

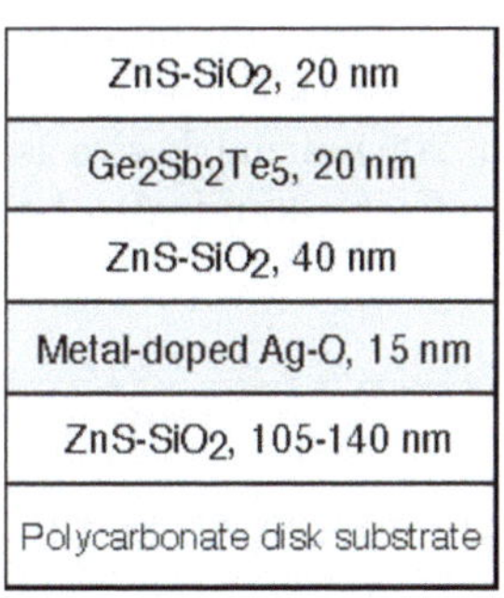

Fig. 8. Super-RENS disk structure

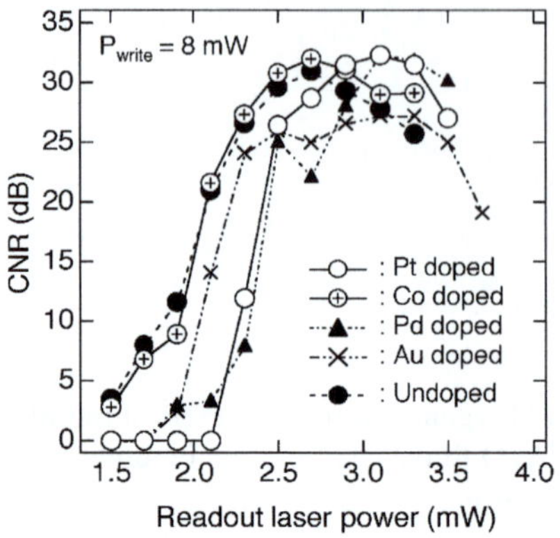

Fig. 9. Relationship between readout laser power and CNR for super-RENS disks with a metal-doped AgO mask layer

with a wavelength (λ) of 635 nm and a numerical aperture (NA) of 0.6. Mark trains with dimension of 200 nm, that is smaller than the resolution limit of 265 nm ($=\lambda/4$ NA), were first recorded at 8 mW and 11 mW ($=P_{\mathrm{write}}$) for a AgO mask and a Ag_2O mask, respectively. Marks were then retrieved by varying the laser power in the range of 1.5–3.9 mW.

Figure 9 shows the carrier-to-noise ratio (CNR) property of the super-RENS disk with a AgO mask. Super-resolutional readout is also achieved for the metal-doped cases, and roughly the same CNR at the maximum of $\approx$ 30 dB was obtained. The minimum laser power necessary for signal retrieval (P_{min}) was 1.5 mW or lower for the undoped case, and it tended to shift slightly towards higher values when Pd, Pt and Au are doped. This indicates that higher laser power (i.e. higher temperature) was necessary for AgO decomposition to achieve super-resolutional readout. The result is in good correspondence with previous optical and XRF studies, though they did not show direct evidence of Ag cluster formation but Ag_2O intermediate. Co doping did not cause a laser power shift, and this shows that its effect on the decomposition was quite small, as is shown in Fig. 7.

Figure 10 shows the CNR property of super-RENS disks with a Ag_2O mask. Probably since the mask is nearly transparent and does not absorb the laser light efficiently, P_{min} became higher compared to that for the AgO mask cases. No distinct shift of P_{min} was recognized, and this indicates that the doping was less effective for control of the Ag_2O decomposition process. It is also shown in the figure that the CNR increased dramatically when the noble metals were doped. The increase of 10–15 dB at 200 nm mark size is attractive for its practical use, and the metal doping effect should also be studied from that point of view.

5 Summary

AgO sputtered film thermally decomposed to Ag_2O at 130–160 °C, and to Ag at $\approx$ 430 °C. Slight noble metal doping ($\approx$ 1 at.%) modified the AgO

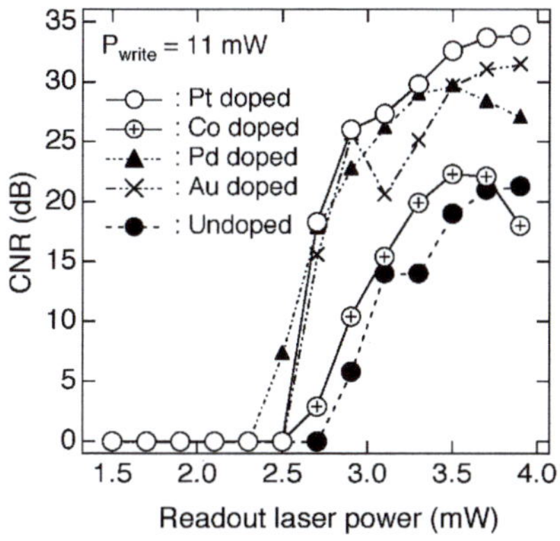

Fig. 10. Relationship between readout laser power and CNR for super-RENS disks with a metal-doped Ag_2O mask layer

film to be thermally stable, and a higher temperature (roughly 200–300 °C) was necessary for the decomposition to Ag_2O. The process showed a close correlation with the laser power necessary for super-resolutional readout in super-RENS disks.

Acknowledgements

The authors would like to thank M. Tateno of Nippon Institute of Technology for his technical assistance. Part of this work was performed at the Nano-Processing Facility (NPF) of AIST.

References

1. J. Tominaga, T. Nakano, N. Atoda: Appl. Phys. Lett. **73**, 2078 (1998) 49, 55
2. H. Fuji, J. Tominaga, L. Men, T. Nakano, H. Katayama, N. Atoda: Jpn. J. Appl. Phys. **39**, 980 (2000) 49, 50, 54, 55
3. D. Büchel, C. Mihalcea, T. Fukaya, N. Atoda, J. Tominaga, T. Kikukawa, H. Fuji: Appl. Phys. Lett. **79**, 620 (2001) 49
4. C. Mihalcea, D. Büchel, N. Atoda, J. Tominaga: J. Am. Chem. Soc. **123**, 7172 (2001) 49
5. D. Büchel, J. Tominaga, T. Fukaya, N. Atoda: J. Magn. Soc. Japan **25**, 240 (2001) 50, 51, 54
6. A. A. Schmidt, J. Offermann, R. Anton: Thin Solid Films **281–282**, 105 (1996) 51
7. J. F. Weaver, G. B. Hoflund: J. Phys. Chem. **98**, 8519 (1994) 51, 52
8. R. Qadeer, F. Ahmad, S. Ikram, A. Munir: J. Chem. Soc. Pak. **21**, 368 (1999) 51, 52
9. M. Epifani, C. Giannini, L. Tapfer, L. Vasanelli, J. Am. Ceram. Soc. **83**, 2385 (2000) 52
10. T. Shima, J. Tominaga: to be published in Thin Solid Films 55
11. K. L. Saenger, C. Cabral, Jr., C. Lavoie, S. M. Rossnagel: J. Appl. Phys. **86**, 6084 (1999) 55

Transient Optical Properties of the Mask Layer for the Super-RENS System

Toshio Fukaya and Dorothea Büchel

Laboratory for Advanced Optical Technology (LAOTECH),
National Institute of Advanced Industrial Science and Technology (AIST),
1-1-1 Higashi, Tsukuba, 305-8562, Japan
t.fukaya@aist.go.jp

Abstract. A super-resolution near-field structure (super-RENS) has an additional mask layer in the usual phase-change optical disk. A thin layer of antimony (Sb) film or a silver oxide (AgOx) layer is used as a mask layer. By focusing a laser beam, a transparent aperture in the Sb layer and a light-scattering center in the AgOx layer are formed transitionally, whose diameters are smaller than that of the laser beam spot. The changed portion can generate an intense optical near field and can be used to record and retrieve small marks beyond the diffraction limit.

The nonlinear optical properties of Sb and AgOx films with protective layers were examined using a pulse laser. Optical switching, their time response and transient spectroscopic change were investigated. The light scattering property and surface-enhanced Raman scattering property of AgOx films were also examined.

A repeated optical switching action can only be realized if the illuminating spot size is confined to very small areas. The time response of the Sb film shows first a rise and then a slow exponential decay. The time response of the AgOx film shows a more complicated decay than the Sb film. Transmittance spectra just after pump irradiation become flat over a wide spectral range in both the Sb and AgOx layers.

Rayleigh scattering and Raman scattering light are extremely enhanced by increasing the input light power, but the fluctuation of the Rayleigh-scattering light intensity does not synchronize with that of the Raman-scattering light intensity.

1 Introduction

Optical data storage systems are a main target for low-cost data media, and many studies have been done on magneto-optical (MO) and phase-change (PC) media. Efforts to increase the recording density in different types of optical data storage systems have been made in many ways. A recording density of more than 100 GB is actually expected in optical disks in the near future; however, it is hard to realize such disks, even using a high numerical aperture lens and a short wavelength laser diode (LD)[1] because of the diffraction limit.

The ultimate optical data storage will use optical near-field recording with the potential of increasing the data storage density up to the terabyte level. *Betzig* et al. [2] proposed the use of a near-field scanning optical microscope and recorded 60 nm marks on MO film. A metal-coated fiber probe, sharpened

J. Tominaga and D. P. Tsai (Eds.): Optical Nanotechnologies,
Topics Appl. Phys. **88**, 59–78 (2003)

to less than the LD wavelength, is used as the recording pickup. The near-field radiated from the fiber tip is effective only in a short region of about 50 nm in size and decays exponentially. The energy transfer rate through the fiber from the LD output to the near-field emission is usually very low, so the distance between the fiber tip and the material surface must be controlled to less than 10 nm with subnanometer accuracy. Therefore, the recording speed and area are limited to a few 10 μm/ s and $\sim 10^4$ μm^2. This technique cannot be applied to actual optical data storage systems because of its low data transfer rate.

Tominaga et al. [3] proposed a method of optical near-field recording to overcome the above difficulties. A thin layer of antimony (Sb) film is used as a nonlinear optical material and inserted in the usual PC optical disk structure. The optical functional material with this new structure, called a super-resolution near-field structure (super-RENS), is the key to this new technique. The transmittance of the Sb film changes drastically with the increase of the input light intensity. When a laser beam is focused on the Sb film, a pinhole with a diameter smaller than that of the laser beam spot is formed transitionally. The size of the light-induced pinhole has not been measured yet directly, but the hole should have a size of the same order as the marks recorded on the PC layer. For instance, PC marks of 60 nm are recorded using 635 nm LD light [4]. An intermediate layer is placed between the Sb and PC layers, and the uniform thickness of this layer sustains the stable condition of near-field interaction between the pinhole and PC layer. As a result, optical near-field recording and readout through this pinhole have been confirmed under high-speed disk rotation in a basic experiment.

We called the super-RENS system using a transparent aperture with Sb film a TA-super-RENS. On the other hand, by using a silver oxide (AgOx) layer instead of a Sb layer, another type of super-RENS was discovered [5]. The AgOx films with various mixtures of silver and oxygen can be produced by reactive RF magnetron sputtering. The transmittance of the AgOx film in the case of Ag_2O film, which is fairly transparent in the as-deposited state, changes drastically with increasing input light intensity. When a laser beam is focused on the Ag_2O film, an opaque spot with a diameter smaller than that of the laser beam spot is formed transitionally. This spot acts as a light-scattering center and produces a very huge near-field interaction [6]. The index change is irreversible by itself in general, but becomes reversible through confining the film between protection layers. The sandwich structure used in the super-RENS has two important roles: the first is to sustain a constant distance between the probe and the PC material, and the second is to prevent the mask component from evaporation. Using the near-field interaction effect though the protective layer, PC marks of less than 100 nm were recorded and retrieved using 635 nm LD light. We called the super-RENS system using a light-scattering center with AgOx film an LSC-super-RENS.

On the other hand, *Büchel* et al. [7] proposed applications of reactively sputtered AgOx thin films as a substrate material for surface-enhanced Raman scattering (SERS). The advantage of these films comes from the highly reproducible and simple fabrication process. The deposited layers developed an increasingly strong SERS activity upon photoactivation at 488 nm. A benzoic acid/2-propanol solution was used to demonstrate that the bonding of molecules to SERS active sites at the surface could be followed by investigating the temporal changes of the corresponding Raman intensities. Furthermore, the laser-induced structural changes in the AgOx layer lead to a fluctuating SERS activity at high laser intensities, which also affects the spectral features of amorphous carbon impurities.

Our objective in this paper is to study the transient optical properties of the Sb and the AgOx films, especially optical switching, its time response and spectral changes, and the light-scattering property of the AgOx film, in a stationary state using a nanosecond pulsed laser arrangement. It is also to study the relationship between the Rayleigh scattering and Raman scattering from the deoxidized silver clusters. And finally it is to get answers to the question, "What happens inside the mask layer?".

2 Experimental

The Sb and AgOx layers were produced by reactive RF magnetron sputtering. To deposit the Sb layer, a pure Sb target was bombarded with Ar gas in the RF plasma. To deposit the AgOx layer, reactive sputtering in an oxygen-containing atmosphere was performed. On increasing the O_2 gas ratio, the film color gradually changes from metallic to dark brown. Samples prepared in this work had a complex refractive index ($n + k\,\mathrm{i}$) of $n = 2.8$ and $k = 0.08$ at 632.8 nm. The index change due to the sputtering gas ratio was referred to in [5]. As protection layers, we used mainly ZnS–SiO_2 (85:15 in at.%), which is widely used as a protection layer for erasable digital versatile disks (DVD-RAM). The refractive index of ZnS–SiO_2 was about $2.25 + 0.01\mathrm{i}$.

The layers were deposited by RF magnetron sputtering with a composite target. The Sb film samples were multilayers of Zn–SiO_2 (20 nm)/Sb (15 nm)/ZnS–SiO_2 (20 nm) deposited on a quartz or a glass substrate. In some cases, we used SiO_2 as a protection layer for the Sb samples. The AgOx film samples were multilayers of ZnS–SiO_2(20 nm)/Ag_2O (15 nm)/ZnS–SiO_2 (20 nm) deposited on a quartz or a glass substrate. These multilayers have the same structures as mask layers for super-RENS applications. For the SERS measurement, the samples were a single layer of AgO (10 s of nm) deposited on a glass substrate.

As a nanosecond pulsed laser, a fundamental 1064 nm light beam and a second-harmonic generated (SHG) 532 nm light beam from a Nd:YAG laser were used at a pulse repetition rate of 10 Hz.

2.1 Transmittance and Reflectance Versus Input Light Power Measurement of Sb Thin Films in a Microscopic Area

The transmittance and reflectance in a microscopic region as a function of input light power were measured by using a single laser beam and a simple method of pumping and probing by itself. Two objective lenses of the Nikon long working distance series were placed face-to-face in a confocal configuration in the optical path. A SHG 532 nm light beam from a Nd:YAG laser was used. The transmitted and reflected light intensities were detected by a Hamamatsu S2281-01 silicon photodiode with a C2719 amplifier. The peak voltages of the detected pulse signals were monitored using an SRS SR250 gated integrator. The total input power at the sample surface was measured using a Sanwa SLP-3000 optical power meter. A Z-scan technique was used to search for moderate conditions of input light intensity to realize a switching action and for the determination of the best focus point.

Figure 1a,b plots the transmittance and reflectance versus the input intensity for each pulse measured at the focal point. Clear switching action can be observed. If the input intensity increases to over about 4.5 nJ/pulse, using the same scale as in Fig. 1, the creation of permanent holes becomes apparent [8,9].

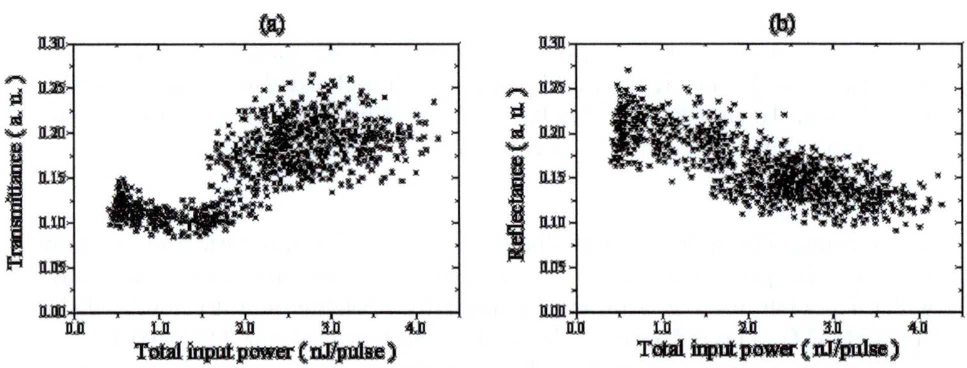

Fig. 1. (**a**) Transmittance versus input intensity. (**b**) Reflectance versus input intensity

2.2 Time Response Properties of Sb Films and AgOx Films

The time response property of the nanometer scale optical switch formed by the Sb film or the AgOx film was examined by using a so-called pump–probe method. For this purpose, a 532 nm nanosecond pulse laser was used as a pump beam. A 442 nm He–Cd laser light was modulated to provide microsecond pulses using a Hoya–Schott acousto-optical modulator and used to probe the samples. The timing and delay of the pump and probe lights were adjusted using two Iwatsu PG-230 pulse generators. The pulse repetition rate of the pump and probe lights was 10 pulses per second, and a pair of pump and probe pulses was selected by using a mechanical shutter and focused on the sample using a microscope. The transmitted light was collected by a counter-facing microscope. The reflected light returned through the objective lens was intercepted by using a beamsplitting cube prism. Two pairs of dispersion prisms and monochromators were used to separate the pump and probe lights in bringing them to the sensors; the transmitted and reflected pump lights were completely blocked. Hamamatsu C5658 avalanche photodiodes with an amplifier module were used as fast and high-sensitive photodetectors and a Sony-Tektronix TDS-3054 digital phosphor oscilloscope was used for recording the time traces of the detected signals. The total input power was evaluated by a Hamamatsu S2281-01 silicon photodiode with a C2719 amplifier, whose sensitivity was compensated using a Gentic ED-100A joule meter with an EDX-1 amplifier as a reference. Figure 2 shows the optical arrangement prepared for the objectives. Without pump irradiation, the clear rectangular signal of the probe light can be observed. With pump irradiation, transitions show a fast changing and slow decaying behavior for both the transmitted and reflected light measurements.

The single-shot-pulse response of Sb film was measured. Figure 3a,b shows traces of the transmitted signal and reflected signal. Without pump irradiation, the clear rectangular signal of the probe light can be observed. The time profile of the initial probe without pump irradiation was obtained by averaging 64 data points. With pump irradiation, transitions with a fast rise and slow decay were observed both in the transmitted light and in the reflected light. With pump irradiation, the reflectance quickly decreased in the rise time of the pump light and slowly recovered. The time profile of the nonlinear characteristics needs to be monitored with a single-shot pump pulse, because the samples are often damaged by pump irradiation. The fluctuations

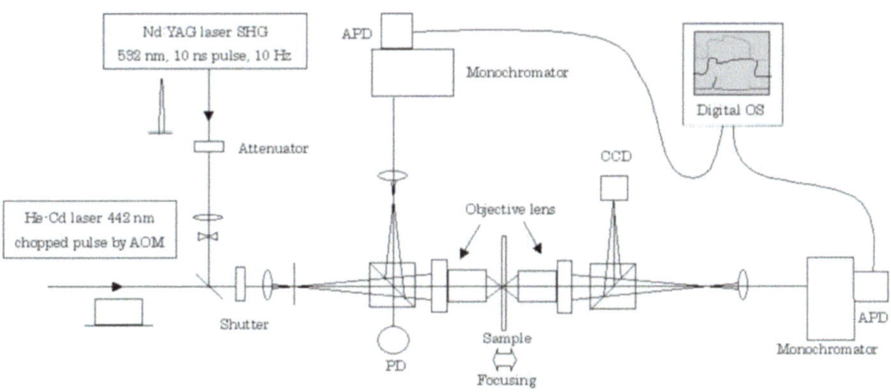

Fig. 2. Schematic diagram of pump–probe measurement system

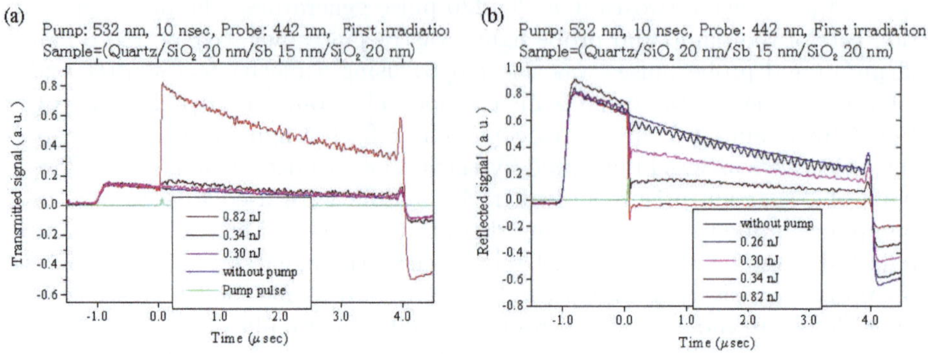

Fig. 3. Single-shot-pulse response of Sb film of (**a**) transmitted signal and (**b**) reflected signal with various pump intensities

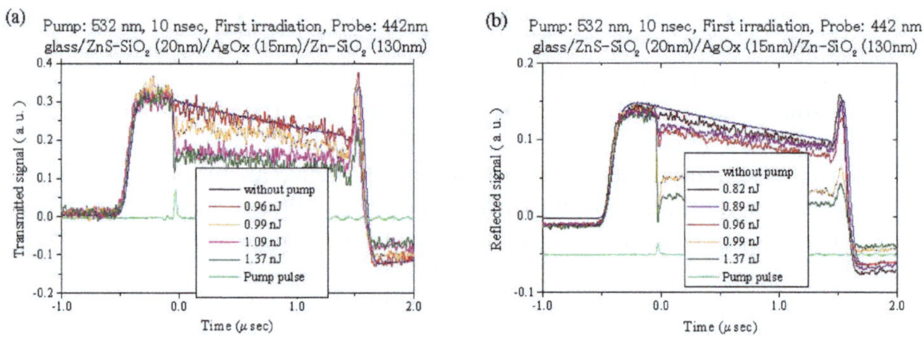

Fig. 4. Single-shot-pulse response of AgOx film of (**a**) transmitted signal and (**b**) reflected signal for various pump intensities

of the recorded signals can be reduced by a numerical filtering technique using a fast Fourier transformation.

The single-shot-pulse response of AgOx film was measured. Figure 4a,b shows examples of these results. The AgOx film shows various responses according to the total input light power of the pump pulse beam. There is some threshold of about 0.8 nJ. When the input power is lower than the threshold, any change is hardly recognizable. First irradiation and repeated irradiations have different responses. A sharp pulse response, which appears in first irradiation, disappears stepwise during the second and third irradiations.

2.3 Spectral Changes of Sb Film and AgOx Film

As shown in the measurements of the time response properties of the Sb film and the AgOx film, transitions caused by the pump irradiation show quick changes in the rise time of the pump light and slow decaying behavior for both the transmitted and reflected light measurements. Spectral changes during the slow decaying period just after the excitation was measured by using white light from a flash lamp (FL) as a probe. The duration time of the FL light was measured to be 600 ns. A fundamental 1064 nm wavelength light beam from a Nd:YAG 10 ns pulsed laser was used to pump the sample. The optical setup of this measurement is shown in Fig. 5.

Because the incoherent white light could not be focused to a small spot, a tapered hole in an aluminum block and a 70 μm diameter pinhole were used to decrease the illumination area. Thus, uniform irradiation for both the pump and probe beams could be achieved and sufficient white light power to analyze the spectrum could be obtained. The total FL light power was weak enough to prevent any irradiation effect from the probe beam. The FL light probe pulse should arrive at the sample just after the pump pulse. The timing and delay of the two lights were adjusted using two Iwatsu PG-230 pulse generators so that the FL light probe pulse arrived about 50 ns after the pump light arrived at the sample. The pulse repetition rate of the pump and probe lights was ten pulses per second, and a pair of pump and probe pulses was selected by using a mechanical shutter. Using a reflective mirror for 1064 nm wavelength light, the pump beam was excluded from reaching a multi-channel analyzer. A Hamamatsu PMA-11 multi-channel analyzer was successfully used to detect the whole spectrum with one-shot irradiation.

Figure 6 shows the FL light spectrum and its transmitted light spectrum through the two 1064 nm reflective mirrors. The 1064 nm reflective mirror also has an opaque region between 380 nm to 430 nm. Using the transmittance spectrum of a quartz base glass as a reference spectrum, we calculate the transmittance spectrum of the samples with and without a pump pulse. Figure 7 shows the transmittance change with and without pump pulse ir-

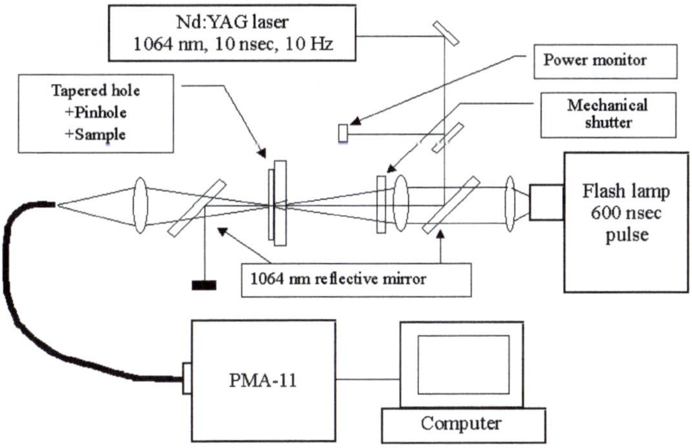

Fig. 5. Optical setup for spectral change measurement

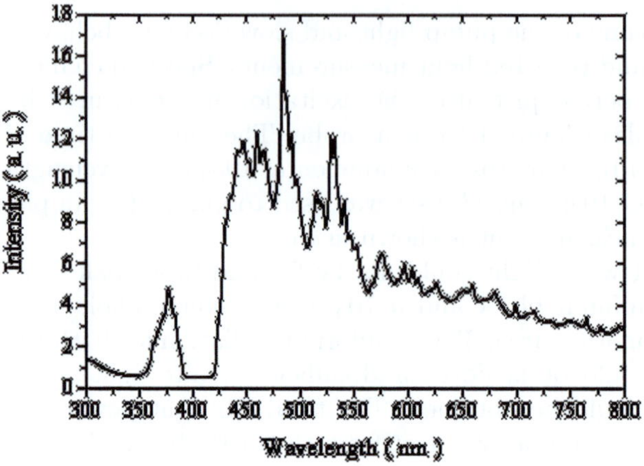

Fig. 6. FL light spectrum through the two 1064 nm reflective mirrors

radiation for the Sb film sample. The Sb film sample has low transmittance, especially at long wavelengths. After the pump irradiation, the transmittance of the Sb sample increases and becomes a flat transmittance spectrum over the whole wavelength range. The pump irradiation area of the Sb sample seemed to be a cavity or a gas vapor. If we increase the pump power more and more, the transmittance will reach almost 100%. Figure 8 shows the transmittance change with and without pump pulse irradiation for the AgOx film sample. The AgOx film sample has a very high transmittance. After the pump irradiation, the transmittance of the AgOx sample decreases and becomes monotonous but lower at low wavelengths. This spectrum does not provide direct evidence of the formation of Ag particles, but is still consistent with that hypothesis. The large downward peak at 532 nm is due to contamination from the SHG light of the pump light.

2.4 Light Scattering Properties of AgOx Films

In the LSC-super-RENS, the light-scattering center acts as an excellent optical near-field probe. AgOx is considered to decompose to silver and oxygen by a photochemical or a thermo-chemical reaction. Illuminating the AgOx film with a focused beam, the illuminated spot becomes bright and brilliant. Scattered light from the illuminated spot is apparently observable by the naked eye if the incident light intensity exceeds some threshold value. Thus the relation between the input light intensity and the scattered light intensity should show some nonlinearity. To study the light scattering property of the AgOx film, the sample was illuminated perpendicularly and the scattered light was monitored under an angle of 45 degrees. Figure 9 shows the schematic of the optical setup for this measurement.

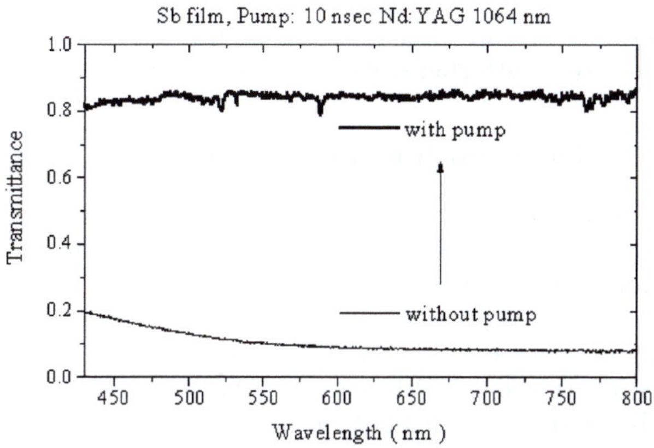

Fig. 7. Transmittance spectrum change by pump pulse irradiation for Sb film

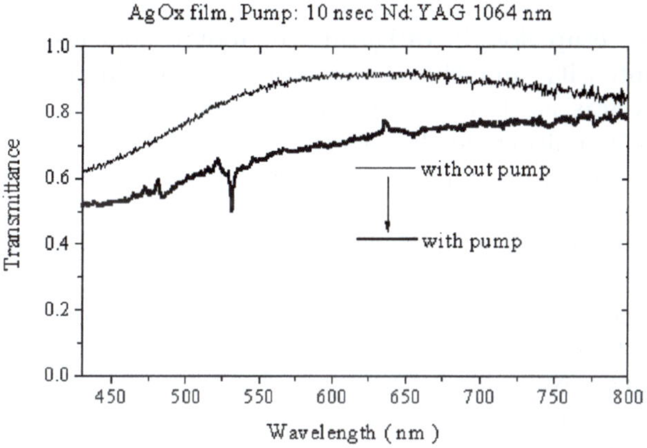

Fig. 8. Transmittance spectrum change by pump pulse irradiation for AgOx film

A SHG 532 nm wavelength light from a Nd:YAG 10 ns pulsed laser is used. The scattered light and total input power were evaluated by two Hamamatsu S2281-01 silicon photodiodes with C2719 amplifiers, whose sensitivity was compensated using a Gentic ED-100 A joule meter with an EDX-1 amplifier as a reference. The peak voltages of the output pulse signals were measured using a Stanford Research Systems SR250 gated integrator. All data were stored for each pulse and the ratio of scattered to incident light intensities individually calculated by use of a computer. The illuminated area was monitored by using a CCD camera and the diameter of the area was estimated to be about 10 μm. Figure 10 shows the plot of the light-scattering efficiency against the total input light power. During measurements, the mean light power was gradually increased by rotating the polarizer of an attenuator. Unfortunately, the switch-on and switch-off positions were not clear, probably because of the instability of the laser power and the large illuminated area. But nevertheless, an extremely high and a very low scattering state can be observed.

2.5 Relation between Rayleigh and Raman Scattering

SERS light considered to be caused by the intereference between surface plasmons and Raman-active molecules which occured on the silver cluster surface. The surface plasmons are also considered to play a very important role in super-RENS for recording and retrieving marks. To get a feel for the generation of surface plasmons, a good way would be to monitor the SERS light versus the Rayleigh scattering light. When the AgOx layer developed an increasingly strong SERS activity upon photoactivation at 488 nm, the laser-induced structural changes in the AgOx layer led to the blinking of the Rayleigh scattering and, at the same time, fluctuating SERS activity at high laser intensities.

Figure 11 shows the setup for investigating temporal changes of the Rayleigh scattering intensity and the corresponding SERS intensity. A backscattering geometry was used. To achieve a rapid measurement, we used a Hamamatsu PMA-11 multi-channel analyzer with a repetition rate of 1 Hz and light from two CW lasers. One is 488 nm wavelength light from a Ar laser for developing SERS activity and exciting Raman scattering. The other is 633 nm wavelength light from a He-Ne laser for monitoring Rayleigh scattering.

A benzoic acid/2-propanol solution was used as a Raman-active material. To block the fundamental 488 nm pump light, a hologram-type notch filter was used. Figure 12a shows an example of the measured obtained spectrum. Figure 12b shows an ordinal Raman spectrum recorded with a Renishaw Ramanscope at a wavelength of 488 nm in the backscattering geometry, and in

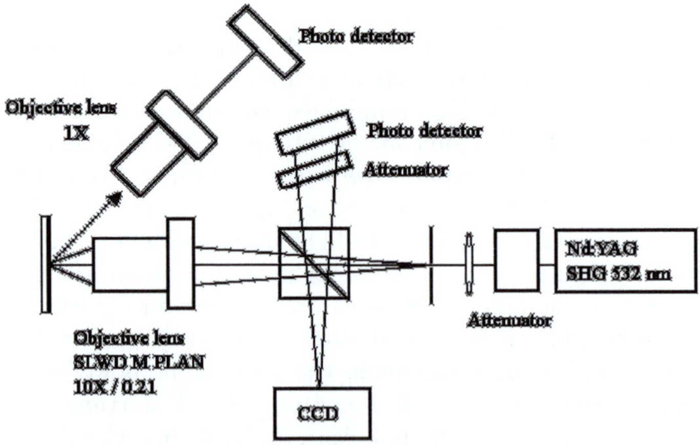

Fig. 9. Optical setup for scattered light measurement

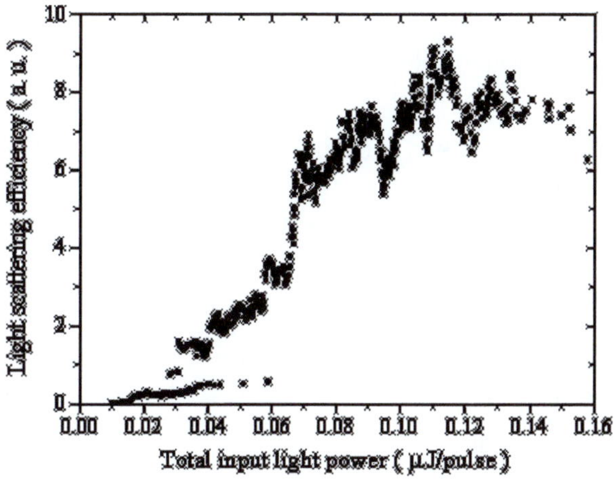

Fig. 10. Light-scattering efficiency versus light power

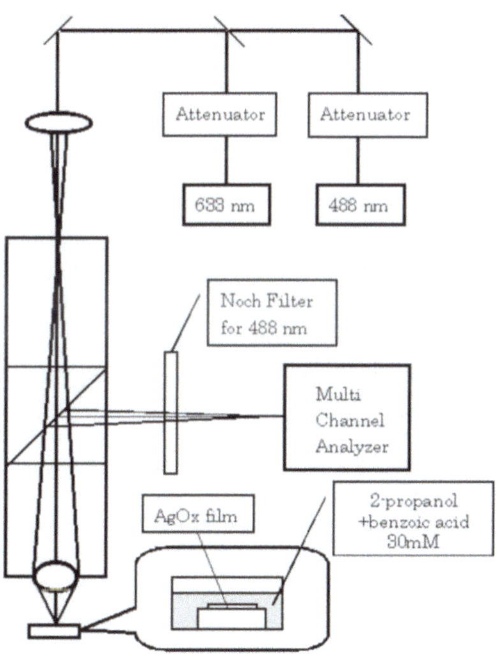

Fig. 11. Setup for simultaneous measurement of surface-enhanced Raman and Rayleigh scattering

this case silver-coated glass was used as a base plate. To monitor the Rayleigh scattering activity, 633 nm wavelength light was guided near the sample surface using an optical fiber. Large luminescence in a range of 580–640 nm of Fig. 12a might be fluorescence from the base glass, but not clear at this moment.

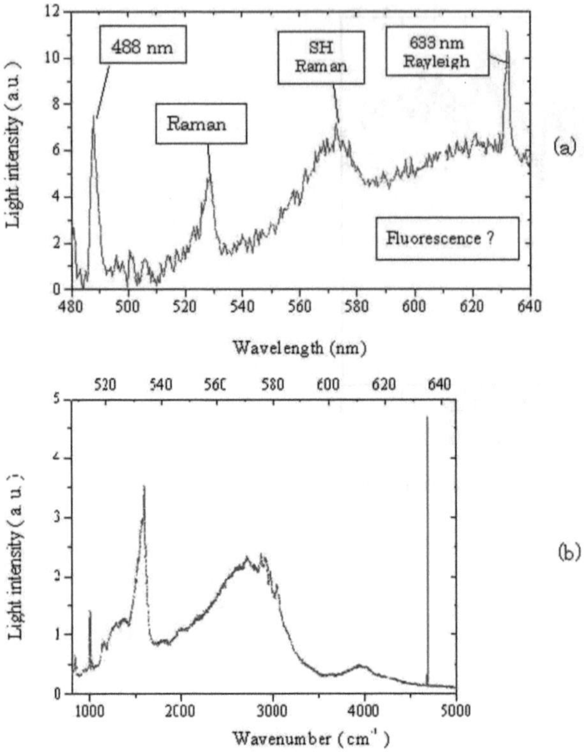

Fig. 12. (**a**) An example of the measured spectrum. (**b**) An ordinal Raman spectrum

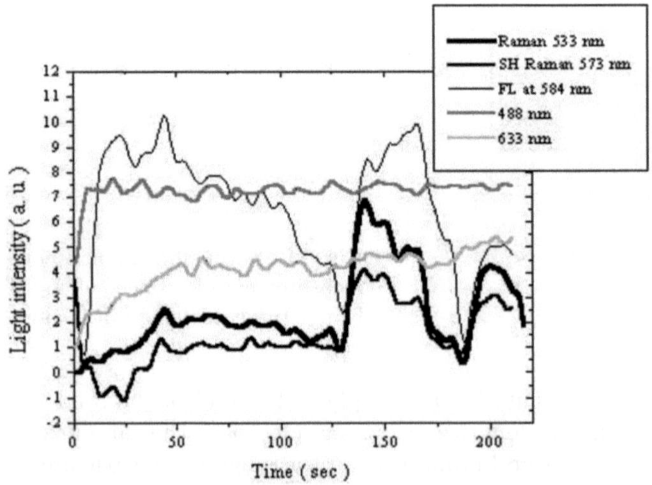

Fig. 13. Time variation plots of the 488 nm stray light, the 633 nm Rayleigh scattering light, the Raman light at 533 nm, the SH Raman light at 573 nm and the fluorescence light at 584 nm

Figure 13 shows time variation plots of the 488 nm wavelength stray light, the 633 nm Rayleigh scattering light, the Raman light at 533 nm, the SH Raman light at 573 nm and the fluorescence light at 584 nm. The 488 nm input light was kept constant and so the 488 nm stray light was almost constant. The 633 nm Rayleigh scattering light increased rapidly and stayed constant as an accumulated average. If we observed it from a limited direction, it was blinking. As the Raman light at 533 nm, the SH Raman light at 573 nm and the fluorescence light at 584 nm were gradually increased, they suddenly began to fluctuate. The fluorescence light showed strange behavior and fluctuated simultaneously with the Raman lights. In any case, the fluctuation of the SERS light intensity has no relation with the blinking of the Rayleigh scattering.

3 Discussion

Tominaga et al. [10] found that sputtered Sb thin films with a thickness of more than 10 nm are usually crystalline. An as-deposited 15.0 nm thick Sb film in a sandwiched structure between protective layers of SiN or SiO_2, etc., is in an intermediate state consisting of amorphous and partly crystalline islands. After half a day at room temperature or after annealing, the film completely changes to the crystalline form. The real and imaginary parts of the refractive index of the crystalline Sb 15.0 nm film at 632.8 nm were measured to be 3.11 and 5.66, respectively and those of amorphous 7.5 nm films were found to be 4.51 and 3.66, respectively. Heating by laser illumination causes a phase transition between the crystalline and amorphous states and this change occurs reversibly by itself through heat diffusion. *Jiang* et al. [11] found that these reversible transitions occur on a timescale of less than 1 μs. A large difference in the refractive index, especially in the imaginary part, causes a large change in the transmittance. These properties might be applied to explain the optical switching behavior in the Sb mask layer.

As a matter of fact, the transmittance and reflectance in the microscopic region as a function of the input light power show clear switching action and the existence of two states, as shown in Fig. 1a,b. But, as will be shown later in Fig. 14, the changes in transmittance and reflectance were monotonous and their variation range was larger than the coverage of the phase-change assumption. As previously described [8,9], if the input intensity exceeds some threshold, the creation of permanent holes becomes apparent. The vanishing of Sb through some crevice in the protection layers may produce a permanent hole. Figure 14 might be a trace of an expansion process of a permanent hole. It is mentioned that melting and further vaporization also seem to happen around the range of 4.5 nJ/pulse in Fig. 1.

From the time response measurement of Sb film, the time response of the reflected probe signal shows a fast decaying time component of about 150 ns, although it is very small and not clear in the transmitted probe signal. The

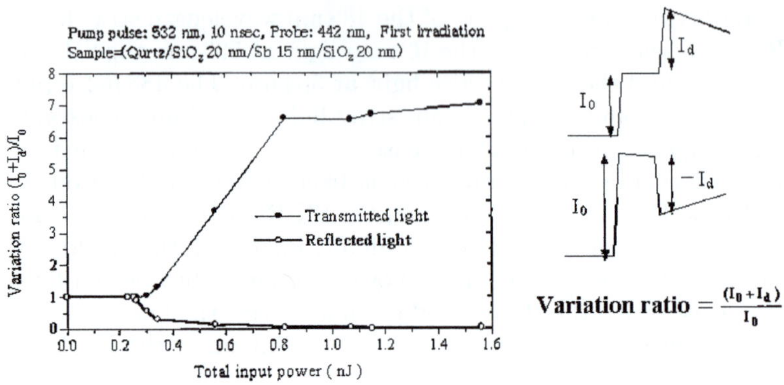

Fig. 14. Variation ratios for transmitted and reflected signals of Sb film plotted against pump light intensity

main decay time is as slow as about 2 μs. It is conjectured that heat from laser illumination causes a large change in the transmitted and reflected light intensities and reversible changes occur by themselves due to heat diffusion. Fast decay times are possibly due to a pure heat diffusion process, and slow decay times indicate a reordering process in the material phase. From the slow decay of the switching action, it is considered that heat from laser illumination causes a large change in the transmittance and the reversible change occurs by itself with heat diffusion.

From the time response measurement of the AgOx film, the time response of the reflected probe signal showed a fast decay-time component of 165 ns, although it is small in the transmitted probe signal. The main decay-time was estimated to be around 1025 ns if we use a fresh sample within a few days after fabrication, as shown in Fig. 4 of [12]. When we measured the same sample after a few months, the fast decay-time component was almost the same, but the main decay-time became three times lager than that of the fresh sample, as shown in Fig. 4.

To compare the result obtained from the time response measurements with the result shown in Fig. 3, we defined a variation ratio as follows; $r = (I_0 + I_\mathrm{d})/I_0$, where I_0 is the initial probe signal intensity and I_d is the variation in the probe signal induced by the pump light irradiation. Figure 14 shows plots of the variation ratio against pump light intensity and both the transmitted and reflected probe signals are calculated. Changes in the reflected light intensity become recognizable above 0.26 nJ/pulse, but cannot be observed in the transmittance measurements below 0.30 nJ/pulse. This fact was checked by repeated measurements in this range. This is explainable, because the reflectance is quite sensitive to the Sb surface state but changes throughout the Sb film are needed for the transmittance change. Above about 0.8 nJ/pulse, the variations seem to be saturated. More details will appear elsewhere [13].

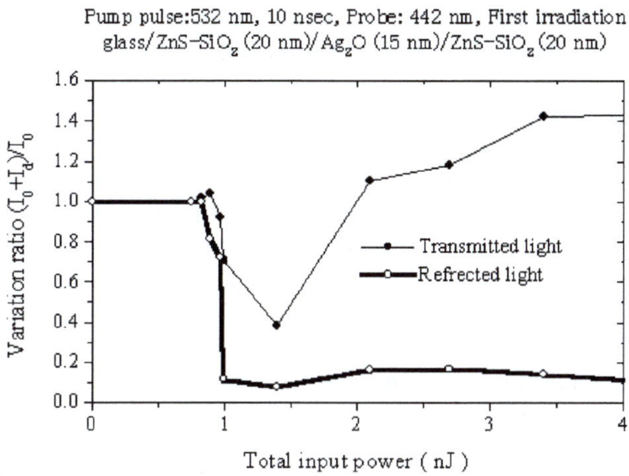

Fig. 15. Variation ratios for transmitted and reflected signals of AgOx film plotted against pump light intensity

Figure 15 shows plots of both the transmitted and reflected probe signals against pump light intensity for the AgOx film. The changes in the transmitted light intensity start at above 0.80 nJ/pulse: first it increases and then it decreases. In the reflectance measurements, on the other hand, the change cannot be observed below 0.85 nJ/pulse. The reflectance decreases monotonically. At about 1.0 nJ/pulse, the variation of the reflected probe light seems to be saturated. At more than 1.5 nJ/pulse, sample bleaching may occur and the transmittance and reflectance approach the values of quartz glass at around 3.0 nJ/pulse. Reciprocal switching characteristics should be obtained only if the total input intensity is in the range between 0.8 to 1.0 nJ/pulse.

When the sample was a super-RENS with a phase-change (PC) layer, glass substrate/ZnS–SiO_2(20 nm)/GeSbTe (20 nm)/Zn–SiO_2(40 nm)/Ag_2O (15 nm)/Zn–SiO_2 (130 nm), the threshold values in the transmitted and reflected light intensity decreased at around 0.15 nJ/pulse. The PC layer has a large absorbance and high reflectance. Heat generated from absorbed light in the PC layer and reflected light at the PC surface promotes the reaction of the Ag_2O layer. A pulse energy of more than 1.0 nJ/pulse is needed to change the phase of the PC layer in case of the pump-probe measurement of sandwiched PC layers. Under 1.0 nJ/pulse, recorded marks can be read out without changing the phase state of the marks, although the super-RENS effect starts above 0.15 nJ/pulse. More details are reported elsewhere [12].

As for the spectral change measurement, a measurement over a wider wavelength range, especially in the shorter wavelength region below 430 nm, would be required for discussion about the formation of the Ag particles. To enable the measurement of shorter wavelengths, replacement of the two reflective mirrors for 1064 nm light from the optics should be considered.

Spectral change within the fast decay-time component would be also interesting. Thus, further investigation using a picosecond pulsed laser is going forward.

Compared with the Sb film, the AgOx film is more variable with respect to the light illumination. Not only under pulse irradiation, but also under weak light illumination, the AgOx film changes step by step. Even if the AgOx film was only weakly illuminated by an Ar-gas CW laser, the scattered light from the AgOx surface gradually increased. Same gradual increasing can also be seen in the low input light region of Fig. 8 before the switching occurred. But to get a large relative intensity of the scattered light, such as after switching, an adequate amount of light power is needed. However, it is difficult to find a clear threshold for switching using fluctuating laser pulses.

Büchel et al. [14] measured the transmission transition of various AgOx layers obtained at a specified oxygen/argon ratio as a function of the sample temperature. The AgO, (Ag_2O+AgO) and Ag_2O layers show large transmission changes between 100 and 150 °C. Figure 16 shows the energy relationship between Ag, Ag_2O and AgO in diagrammatic form. Ag_2O is most stable and AgO is considered to be gradually decomposed into O_2 and Ag_2O. Ag_2O is decomposed into Ag and O_2 at over 160 °C by light irradiation and the rise in heat. Photo-thermally decomposed Ag and O_2 might be recombined to make Ag_2O if they are confined between protective layers.

It is very natural to think that the Rayleigh scattering light is enhanced by the progress of deoxidation of AgOx and the growth of silver clusters. The blinking of the Rayleigh scattering light is thought to be the effect of random interference speckle, and a small change in the internal phase condition of silver clusters will cause constructive or destructive interference change. Actually, we can observe the moving speckle pattern on the room wall concurrently with the start of blinking at 488 nm input light wavelength. The photoactivation to develop an increasingly strong SERS activity and the generation of a surface plasmon should have quite different processes compared to the Rayleigh scattering.

To understand or to get a feel of a surface plasmon, a finite difference time domain (FDTD) calculation was done. A plane wave light source at the top segment line was assumed with a Y-polarizing direction. To avoid an

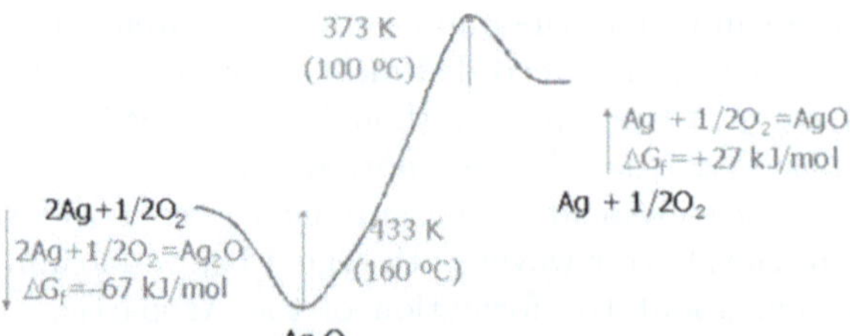

Fig. 16. Energy diagram of various silver oxides

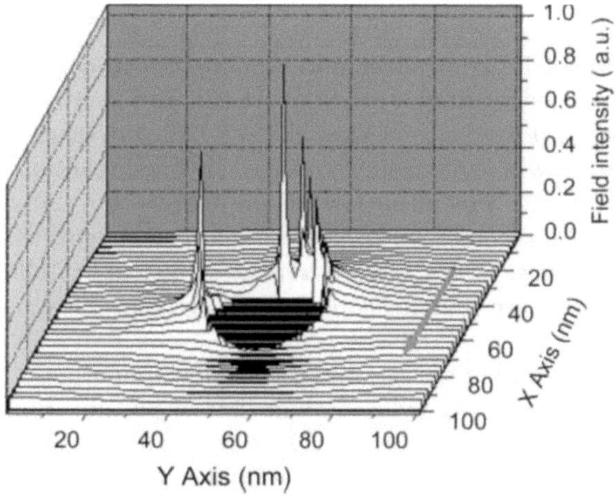

Fig. 17. FDTD simulation of the near-field intensity at the silver particle edge

initial irregular response, the amplitude was set to be gradually increasing and then become constant. We divided the area into 1000×1000 segments, each segment being $1\,\text{nm}^2$. A silver particle with a diameter of 30 nm in the 2-propanol solution was assumed.

The displayed area in Fig. 17 is 100 segments square and it is interesting that surface plasmons appear around the silver particle and have greater amplitude at the cratering surface. This silver particle has a rough surface because of segmentation. It looks like a rock and waves at the sea coast. From this image, the "hot spot" might be explainable. The name of "hot spot" refers to a special spot on the silver surface and SERS is strongly active at this spot. In some cratering, where the surface bends inside, the surface will be rich in positively charged ions and Raman-active molecules with negatively charged ions will be absorbed. Also, cratering acts as a plasmon-active site.

4 Conclusion

In conclusion, we have confirmed the formation of a nanometer-scale optical switch in sandwiched Sb film using 10 ns pulsed laser light irradiation. It is important to note that repeated optical switching action can only be achieved if the illuminating spot size is confined to very small areas. When larger areas are illuminated, the sandwiched layers are destroyed and the switching properties disappear. Melting and furthermore vaporization will happen inside the Sb layer. The decay mechanism is related to heat diffusion inside the films. If heat can quickly be transported out of the illuminated region, the films recover before being destroyed, possibly due to different thermal expansion coefficients or gas bubble formation.

The AgOx layers are made from various mixtures of AgOx and O_2. According to the energy diagram of Fig. 16, silver and AgOx tend to change toward Ag_2O. The AgOx layer as-deposited state is uncertain and very sensitive to light and heat exposure. If we consider that the AgOx layer consists of Ag_2O, optical switching action will come from a thermo-chemical reaction: deoxidation to Ag and O_2. To realize a reciprocal switching, photo-thermally decomposed Ag and O_2 should be recombined to make Ag_2O and, of course, they are confined between protective layers. Clearly repeated optical switching action is hardly realized and there remains some hysteresis, especially in the initial stage.

The time response of Sb film shows first a rise and then a slow monotonous decay. As for the AgOx film, its sensitivity is lower than that of Sb film. But in the case of the super-RENS structure, in which AgOx is used with a phase-change material such as a GeSbTe, it has a very low threshold around 0.35 nJ. The chemical reaction of AgOx associated with light irradiation seems milder than that of the Sb bubbling. The AgOx layer used with the PC layer shows higher sensitivity and is more stable than the Sb layer. In this case, light is mainly absorbed at the PC layer and heat from the PC layer causes the thermo-chemical reaction of AgOx. The AgOx film might be less useful than the Sb film as a micro-optical switch. But, we consider that the LSC-super-RENS is more promising than the TA-super-RENS.

As the laser-scattering center, the silver cluster itself scatters a high percentage of the input light and produces random interference speckle. Some surface plasmon effect occurring on the silver cluster surface should increase this Rayleigh scattering light. Basic aspect about the generation of surface plasmon was obtained by FDTD calculation. But many important factors, which are necessary to reveal the mechanism and the influence of the surface plasmon in super-RENS for recording and retrieving marks, remain unclear. These are, for example, the relations between the surface plasmon and the Rayleigh scattering, the size and shape of the silver clusters and their movements, etc.

Since the purpose of the super-RENS is to record and retrieve marks with dimensions below the diffraction limit, the protective layered structure seems to be an ideal arrangement in this respect. Third-order nonlinear optical materials usually show intensity-dependent refractive indices, and the relationship between transmittance and input light power is almost linear. The optical nonlinearity of the sandwiched Sb film and AgOx film differs from that of ordinary nonlinear optical materials, and the thermally induced phase change or photo-chemical reaction in the layer turns the structure into an excellent optical near-field generating element. Without the protective layers, the mask layers do not show repeated switching properties. Therefore, the layered structure is the key element of the super-RENS technique and further improvements have to address the mask layer materials and also the protective layer materials and their arrangement.

References

1. M. Kato, Y. Kitaoka, K. Yamamoto, K. Mizuuchi: Tech. Dig. Joint MORIS/ISOM '97, Yamagata, Japan, 27. Oct. 1997 (Business Center Acad. Soc. Jpn., Tokyo 1997) p. 46 59
2. E. Betzig, J. Trautman: Science **257**, 189 (1992) 59
3. J. Tominaga, T. Nakano, N. Atoda: Appl. Phys. Lett. **73**, 2078 (1998) 60
4. J. Tominaga, T. Kikukawa, M. Takahashi, K. Kato, T. Aoi: Jpn. J. Appl. Phys. **36**, 3598 (1997) 60
5. H. Fuji, J. Tominaga, T. Nakano, N. Atoda, H. Katayama: ISOM/ODS '99, Hawaii, 1999, SPIE **3864**, 33 (1999), paper TuD29 60, 61
6. H. Fuji, J. Tominaga, T. Nakano, N. Atoda, H. Katayama: Jpn. J. Appl. Phys. **39**, 980 (2000) 60
7. D. Büchel, C. Mihalcea, T. Fukaya, N. Atoda, J. Tominaga, T. Kikukawa, H. Fuji: Appl. Phys. Lett. **79**, 620 (2001) 61
8. T. Fukaya, J. Tominaga, T. Nakano, N. Atoda: Appl. Phys. Lett. **75**, 3114 (1999) 62, 71
9. T. Fukaya, J. Tominaga, N. Atoda: Proc. SPIE **4268**, 79 (2001) 62, 71
10. J. Tominaga, H. Fuji, A. Sato, T. Nakano, T. Fukaya, N. Atoda: Jpn. J. Appl. Phys. **38**, 4089 (1999) 71
11. F. Jiang, M. Okuda: Jpn. J. Appl. Phys. **80**, 97 (1991) 71
12. T. Fukaya, D. Büchel, S. Shinbori, J. Tominaga, N. Atoda, D. P. Tsai, W. C. Lin: J. Appl. Phys. **89**, 6139 (2001) 72, 73
13. T. Fukaya, J. Tominaga, T. Nakano, N. Atoda: Proc. SPIE **4085**, 197 (2000) 72
14. D. Büchel, J. Tominaga, T. Fukaya, N. Atoda: J. Magn. Soc. Japan **25**, 240 (2001) 74

Index

A Thermal Lithography Technique Using a Minute Heat Spot of a Laser Beam for 100 nm Dimension Fabrication

Masashi Kuwahara[1], Christophe Mihalcea[1], Nobufumi Atoda[1], Junji Tominaga[1], Hiroshi Fuji[2], and Takashi Kikukawa[3]

[1] Laboratory for Advanced Optical Technology (LAOTECH), National Institute of Advanced Industrial Science and Technology (AIST), Central 4, 1-1-1 Higashi, Tsukuba, Ibaraki, 305-8562, Japan
kuwaco-kuwahara@aist.go.jp

[2] Advanced Technology Research Laboratories, Sharp Corp., 2613-1, Ichinomoto, Tenri, Nara, 632-8567, Japan

[3] Information Technology Research Center, TDK Corp., TDK Chikumagawa 1st technical center, 462-1, Otai, Saku, Nagano, 385-0009, Japan

Abstract. We propose a new lithography technique called "thermal lithography" for patterning fine structures far beyond the diffraction limit. A focused laser spot was used to produce a spatially confined hot area in a photoresist film. This technique enabled us to succeed in fabricating lines and dots with 100 nm dimensions in the photoresist film. The dimensions of the produced patterns correspond to about one fifth of the diffraction limit of 530 nm defined by our optical setup. We will apply this technique to the mastering process for optical ROM disks as a low-cost lithography technique.

1 Introduction

In the optical ROM data-storage mastering process [1], a lithography technique for producing 100 nm dimension pits will be necessary for ultrahigh density recording in the future. It is well known that for lithography techniques, shorter wavelengths [2,3,4] or electron beams [5,6,7] have been strong candidates and have been studied at several laboratories. However, shortening the wavelength faces the difficulty of developing improved light sources as well as optical components, while the electron beam technique needs vacuum installations and high-voltage power supplies for handling the electron beam, thus requiring large-scale equipment. These problems turn these techniques into high-cost processes for the optical ROM disk mastering process.

Recently, we demonstrated that a super-resolution near-field structure (super-RENS) [8] technique can be used to produce narrow grooves of 100 nm dimension in a photoresist film by a simple process unsing the optical near-field [9,10,11]. The Technique was also successful for fabricating such narrow grooves at a high speed of about 10^5 times that of the conventional

J. Tominaga and D. P. Tsai (Eds.): Optical Nanotechnologies,
Topics Appl. Phys. **88**, 79–87 (2003)

scanning near-field optical microscope (SNOM) fabrication technique. The super-RENS technique is a practical possibility as a low-cost technique for sub-wavelength fabrication because it permits the use of visible light sources and conventional optics. Nevertheless, we found that the photoresist film was damaged by the laser heating and remained on the sample surface after the development process.

2 Technique

In this paper, we propose a new lithography technique called the "thermal lithography" technique because it uses laser heating. In this technique, we heat confined areas of a reversible photoresist far beyond the diffraction limit. That is, adjusting the applied laser power and the disk rotation speed, the heated region in the photoresist film can be confined to an area with dimensions far beyond the diffraction limit. A reversible photoresist, which is intended for image-reversal patterning through an annealing and a light-exposure process, was selected for this purpose. A conventional phase-change material was also selected to act as a light-absorption layer and to generate the minute heated area. Thermal cross-linking occurs in the heated photoresist portion only in the highest-temperature region inside the focused laser beam and small structures are formed. A semiconductor laser source was also used for heating the photoresist film as in the super-RENS technique; however, only one light source is required for lithography in the new technique. It is shown below that this "thermal lithography" technique has large potential for the low-cost production of pits in disk mastering.

Figure 1 shows the distribution of the light intensity and the temperature inside a focused laser spot. The light intensity distribution inside the focused laser spot has a nearly Gaussian cross-sectional profile. This distribution is able to generate the effective temperature area at the focused spot center. Here, the effective temperature area means the temperature in the area is enough or more for inducing the thermal cross-linking. Moreover, adjusting the laser irradiation time and power allows control of the extent of the heated area inside the illuminated material. Therefore, it is possible to confine a chemical or physical reaction inside this heated area. Indeed, it is possible to make minute recorded marks with dimensions of less than 100 nm at the phase-change recording layer [8]. In this study we made the chemical reaction known as the thermal cross-linking reaction occur in the reversible photoresist film inside the minute area. The reacted portion of the photoresist appears on the sample surface after the development process and forms convex structures with a sub-wavelength scale.

Figure 2 shows the process for the production of fine structure on a substrate. First, the spin-coated photoresist film, which is of the image-reversal type, is flood-exposed with blue light from a mercury lamp ($\lambda = 365$ nm), creating acid. After that, a red laser beam is irradiated onto an absorption

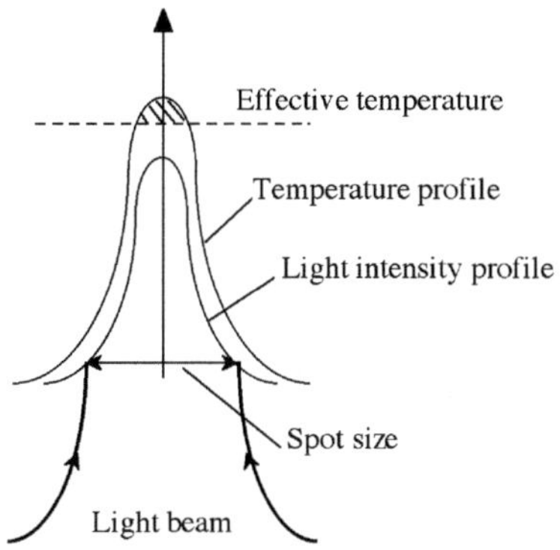

Fig. 1. Light intensity and temperature distribution inside a focused laser spot

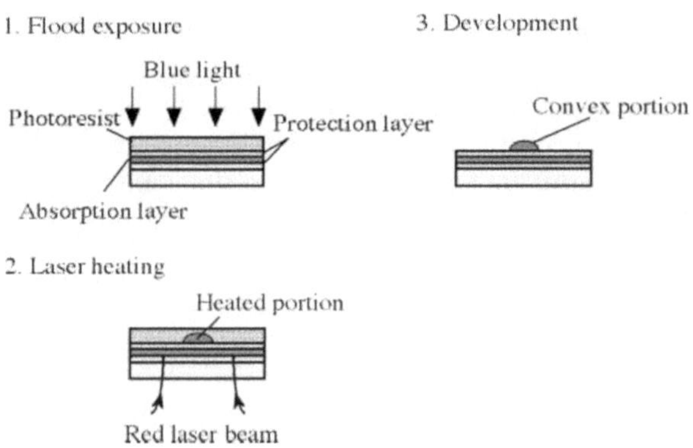

Fig. 2. Process for production of fine structure at thermal lithography

layer and transformed into heat at the layer. Then the generated heat propagates to the photoresist film and causes a cross-linking reaction in it. Here the created acid works as a catalyst through the cross-linking reaction. The cross-linked photoresist portion is no longer soluble in the development solution. Consequently, convex fine structures are formed on the sample surface after development.

The cross-sectional structure of our sample is shown in Fig. 3. The substrates were commercial optical disks made of polycarbonate with lands and grooves with 1.2 μm pitch width. A ZnS–SiO_2 (top layer)/$Ge_2Sb_2Te_5$ (GST)/ZnS–SiO_2 (bottom layer) multilayer structure was deposited on the disk substrate by RF-magnetron sputtering. The thicknesses were 20 nm, 15 nm and 200 nm, respectively. The GST film is in general use as a material for recording layers and works as a laser-light absorber and generates heat in this study. The ZnS–SiO_2 films are protection layers for the polycarbonate and the GST film against heat damage and oxidation. After the depositions, an hexa-methyl-disilazane (HMDS) treatment was carried out by a N_2 bubbling method so as to improve the adhesion of the photoresist

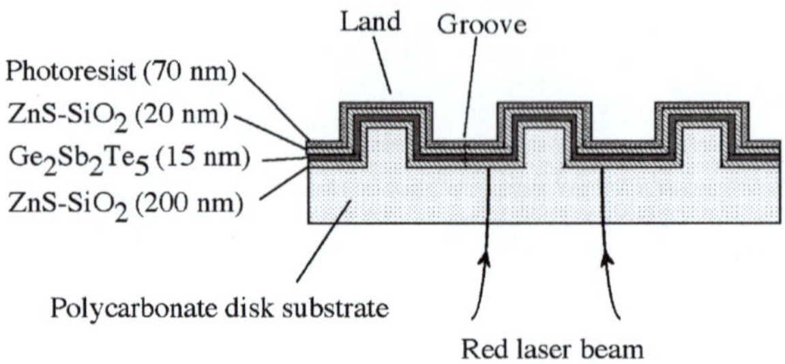

Fig. 3. Cross-sectional view of sample disk

film and avoid exfoliation of the film during the development process. The photoresist (AZ5214E, image-reversal patterning photoresist, Clariant Co.) was spin-coated on the ZnS–SiO_2 top layer after diluting it with a commercial thinner solution, and its thickness was about 70 nm. After coating, all sample disks were prebaked for one minute at 90 °C and flood-exposed by blue light (mercury lamp, $\lambda = 365$ nm). A semiconductor laser with 635 nm wavelength was used as the light source for generating the heat because the combination of that light wavelength and the GST film is a typical setup for phase-change recording and enables the generation of a minute hot area.

3 Experiment

Figure 4 shows the experimental setup. The sample disk was set on an optical disk drive tester (DDU-1000, Pulstec Industrial Co., LTD.). The disk was rotated at a constant linear velocity (CLV) of 6.0 m/ s throughout our experiments. A semiconductor laser irradiated the disk from the polycarbonate substrate side and was automatically focused on the GST film through an objective lens with a numerical aperture of 0.6. The diameter of the laser spot and the diffraction limit for our optics were about $\lambda/(\mathrm{NA}) = 1.06$ μm and $\lambda/(2\mathrm{NA}) = 530$ nm, respectively. Continuous and pulse modes of the laser were used for the formation of lines and dots respectively. The laser power in the continuous mode was changed from 2.0 mW to 5.0 mW at 0.5 mW intervals. The power in the pulse mode was changed from 8.0 mW to 15.0 mW at 1.0 mW intervals. All experiments were carried out only on land areas of the disk because it is easy to obtain topographies on the lands rather than on the grooves when using an atomic force microscope (AFM). The samples were developed with an organic alkali solution (NMD-W, Tokyo Ohka Kogyo Co., LTD.) for 30 s and then rinsed in pure water for 30 s, both processes at room temperature. After that, post-baking was carried out at 110 °C for 60 s. AFM was used in the tapping mode for observation of the topographies and the determination of the dimensions of the fabricated structures.

Figure 5 shows the topography of a convex line observed on a land. The laser in continuous mode was irradiated for 10 s, corresponding to about 400

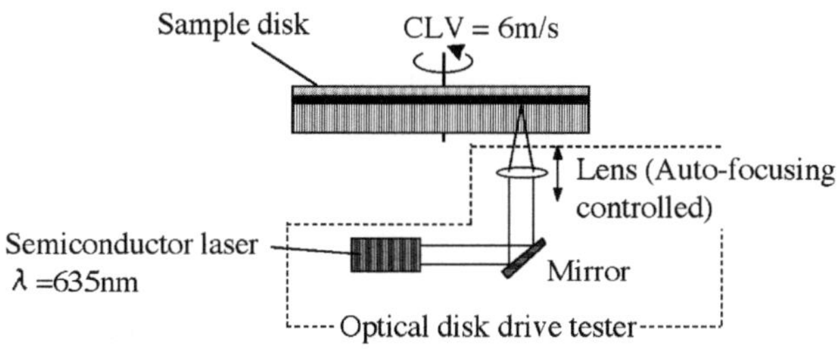

Fig. 4. Experimental setup. The sample disk is mounted onto a disk drive tester with an auto-focusing system

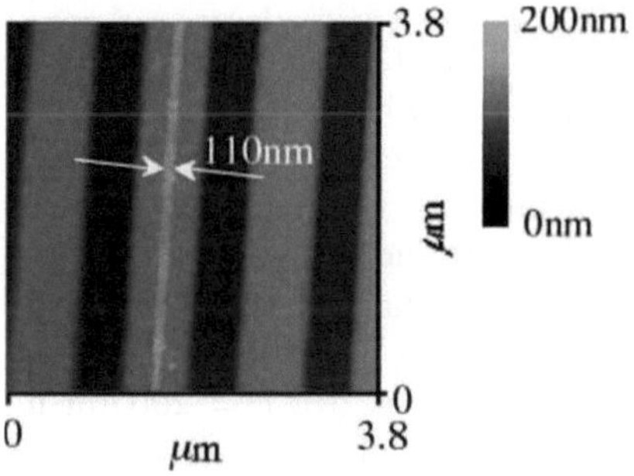

Fig. 5. Topography of a narrow line with 110 nm width produced by continuous laser illumination

disk rotations, and the laser power was 4.0 mW. The linewidth was approximately 110 nm and far narrower than the laser spot size ($d = 1.06$ μm) and also the diffraction limit ($d = 530$ nm). The height was about 35 nm and about half of the thickness of the spin-coated photoresist film. Our technique can confine the extension of the thermal cross-linking reaction within dimensions of 110 nm width and 35 nm height. On the other hand, no line structures could be observed for powers of less than 3.5 mW and more than 4.5 mW on the land area. It is clear that the temperature inside the photoresist film at less than 3.5 mW of the laser power is not sufficient for stimulating the cross-linking reaction. At more than 4.5 mW, it is thought that the photoresist was evaporated or removed from the land areas because residual photoresist annealed by laser light was observed on the edge of lands or grooves in the AFM images.

Figure 6 shows an AFM image of a dot structure fabricated by pulsed laser irradiation on a land area during one disk rotation. Typical dot dimensions were 105 nm in diameter and 20 nm in height. The pulse conditions were 6 MHz frequency, 30% pulse duty ratio and a power of 14 mW, respectively. The laser-spot running length (l) on the disk surface was calculated from the pulse frequency (f), the duty ratio (α) and the disk rotation speed (v) to be $l = \alpha v / f = 300$ nm. However, the obtained dot size is far smaller than this distance. Obviously, at the beginning of the power-on cycle, the generated temperature in the photoresist film is not sufficient to stimulate the cross-linking reaction. The heat accumulates gradually inside the photoresist film

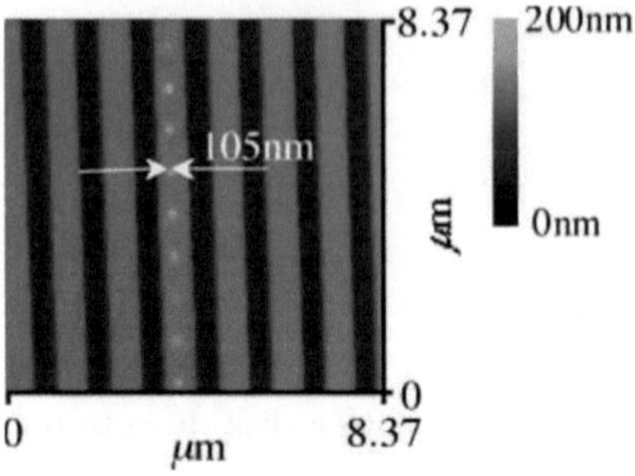

Fig. 6. Dot image produced by pulse laser illumination with a frequency 6 MHz and a duty ratio of 30%

as the laser spot continues to heat the surface. The threshold temperature level to induce the cross-linking reaction is expected to be located at the end of the exposed area. Consequently, dot sizes were achieved with a diameter of 105 nm, which is one fifth of the diffraction limit.

Another experiment was carried out for the production of smaller dots than those shown in Fig. 6. For this purpose, we changed the duty ratio from 30% to 20%, while the pulse frequency and the power were fixed at 6 MHz and 14 mW. The laser running distance was $l = \alpha v/f = 200$ nm in this experiment. Figure 7 shows the AFM topography of the dot structures generated under these conditions. The obtained dot dimensions were typically 80 nm in diameter and 15 nm in height. This diameter corresponds to less than one sixth of the the diffraction limit of the experimental optics. However, the ratio of the dot diameters between those in Fig. 6 and Fig. 7 was not proportional to the reduction of duty ratio or the laser-spot running distance. Nevertheless, changing the duty ratio seems to control the dot dimensions under these conditions.

Figure 8 also shows an AFM topography image of dot structures generated under the conditions of 9 MHz (frequency), 20% (duty ratio) and 15 mW (laser power). Typical dot dimensions were 150 nm in diameter and 20 nm in height. This size was larger than that of the previous experimental results for 6 MHz, although the calculated laser-spot running length was 130 nm, which was shorter than that of the previous experiments. Here the cool-down time

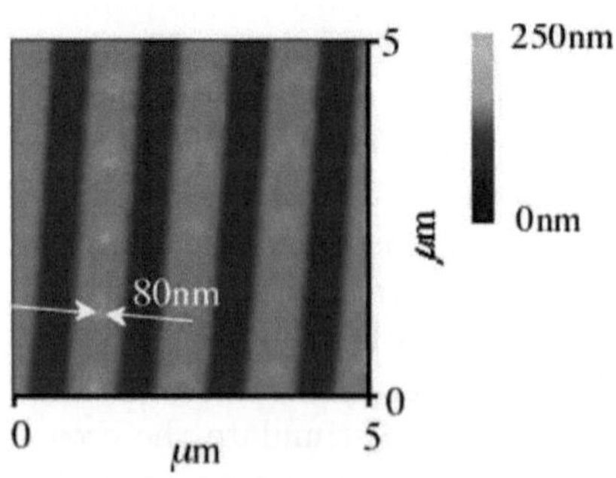

Fig. 7. Dot image produced by pulsed laser illumination with a frequency 6 MHz and a duty ratio of 20%

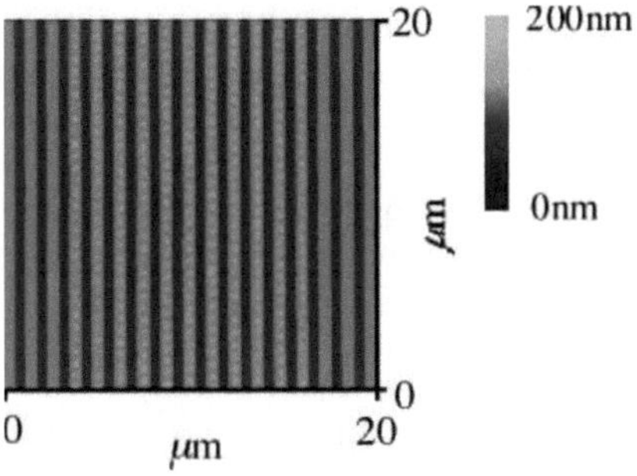

Fig. 8. Dot image produed by pulsed laser illumination with a frequency 9 MHz and a duty ratio of 20%

between the pulses seems to be too short to produce individual spots with smaller dimensions. Further investigations are needed to clarify the relation between pulse frequency, duty ratio and applied laser power to control the dot size and interval spaces between the dots for the fabrication of various pits and to elucidate the dot-formation mechanism.

We also have tried to produce lines or dots on flat disk substrates made of fused silica without lands and grooves, because pit production on a flat substrate is necessary for the optical ROM mastering process. Nevertheless, we could not produce any fine structure with sub-wavelength dimensions by either continuous or pulsed laser mode on silica substrates. The heat inside the photoresist film cannot be confined in the sub-wavelength area because the thermal conductivity of the fused silica with the value of 1.4 W/(m · K) [12] is seven times that of the polycarbonate with the value of 0.19–0.15 W/(m · K) [13,14]. That is, the generated heat on the silica substrate can diffuse more easily than that on the polycarbonate substrate, and confining the heat inside the minute area is therefore more difficult. Computer simulation is thought to be a powerful tool for the thermal analysis in each material and for considering the configuration and shape of the substrates.

We have demonstrated the fabrication of lines and dots with 100 nm dimensions far beyond the diffraction limit of the optical system used by applying a "thermal lithography" technique. We have found that the line formation was very sensitive to the laser power and that the dot dimensions could be controlled by the duty ratio of the laser pulses. We believe that this technique has large potential as a low-cost technology for the disk mastering process in ultrahigh-density optical ROM technology.

References

1. A. B. Marchant: *Optical Recording, A Technical Overview* (Addison Wesley, Massachusetts 1990) 79
2. Y. Kaneda, S. Kubota, H. Yamatsu, M. Furuki, K. Kurokawa, T. Kashiwagi: Jpn. J. Appl. Phys. **37**, 2125–2129 (1998) 79
3. M. Takeda, M. Furuki, H. Yamatsu, T. Kashiwagi, Y. Aki, A. Suzuki, K. Kondo, M. Oka, S. Kubota: Jpn. J. Appl. Phys. **38**, 1837–1838 (1999) 79

4. M. Takeda, M. Furuki, T. Ishimoto, K. Kondo, M. Yamamoto, S. Kubota: Jpn. J. Appl. Phys. **39**, 797–799 (2000) 79
5. T. Imai, N. Shida, K. Suda, T. Higuchi: Jpn. J. Appl. Phys. **40**, 1661–1665 (2001) 79
6. G. Gartwright, G. Reynold: Int. Sympos. Optical Memory, Chitose, Hokkaido, Japan, Sept. 5–8, 2000, Tech. Dig. (Business Center Acad. Soc. Jpn., Tokyo 2000) pp. 218–219 79
7. M. Katsumura, H. Kitahara, M. Ogasawara, Y. Kojima, Y. Wada, T. Iida, F. Yokogawa: Jpn. J. Appl. Phys. **40**, 1653–1660 (2001) 79
8. J. Tominaga, T. Nakano, N. Atoda: Appl. Phys. Lett. **73**, 2078–2080 (1998) 79, 80
9. M. Kuwahara, T. Nakano, J. Tominaga, M. B. Lee, N. Atoda: Jpn. J. Appl. Phys. **38**, L1079–L1081 (1999) 79
10. M. Kuwahara, T. Nakano, J. Tominaga, M. B. Lee, N. Atoda: Microelectron. Eng. **53**, 535–538 (2000) 79
11. M. Kuwahara, T. Nakano, C. Mihalcea, T. Shima, J. H. Kim, J. Tominaga, N. Atoda: Microelectron. Eng. **57–58**, 883–890 (2001) 79
12. D. R. Lide (Ed.): *Handbook of Chemistry and Physics*, 78th edn. (CRC Press, Boca Raton 1997) Sect. 12, pp. 197 85
13. *Vacuum Handbook* (ULVAC Corp. Center, Tokyo 1992) pp. 106 85
14. T. W. McDaniel: J. Magn. Soc. Jpn. **23**, 251–256 (1999) 85

New Structures of the Super-Resolution Near-Field Phase-Change Optical Disk and a New Mask-Layer Material

Lu Ping Shi and Tow Chong Chong

Data Storage Institute (DSI),
DSI Building, 5 Engineering Drive 1, 10 Kent Ridge Crescent, 117608 Singapore
SHI_Luping@dsi.a-star.edu.sg

Abstract. The removable and rewritable high-density phase-change optical disk is attractive for high-definition television (HDTV), digital TV and internet downloading. Super-resolution near-field technology is one of the most promising technologies for achieving ultra-high density recording and is considered a more feasible way of near-field optical recording with simpler recording-head design, less mechanical damage, no contamination and higher recording speed. Developing new structures and searching for new mask materials are two important issues.

1 Introduction

Removable and rewritable high-density and low-cost media are attractive for high-definition television (HDTV), digital movies TV and internet download. Phase-change (PC) media [1,2,3,4] and magneto-optical (MO) [5,6,7,8] are the two most promising technologies. But both of the optical-recording technologies employ a laser beam. The laser-beam spot size is limited by the diffraction limit, which is dependent on the wavelength λ and the numerical aperture (NA) of the objective lens. For example, a digital versatile disk (DVD) system that can store 4.7 GB data uses a laser diode (LD) with a wavelength of 635–650 nm and a 0.6 NA, which restrict the spot size to about 0.8 μm [9]. An easier way to reduce the spot size and increase the data density is to utilize a short-wavelength LD and a higher-NA objective lens.

Philips invented the digital video recording (DVR) system which allows the storage of 22.5 GB on a single-layer 12 cm disk by utilizing a blue laser (405 nm) and a dual-objective lens with $\mathrm{NA} = 0.85$ [10]. *Matsushita* proposed a dual-layered disk to double the capacity [11]. Recently Matsushita demonstrated a 50 GB dual-layer phase-change DVR with a transmittance-balanced structure [12].

For MO recording, useful technologies are super-resolution techniques such as MSR (magnetic super-resolution), PSR (premastered super-resolution), MAMMOS (magnetic amplifying magneto-optical system), and CoSR (Co-super-resolution). All of these technologies have been developed to generate a small aperture in high-density optical disks [13,14,15,16,17,18,19].

J. Tominaga and D.P. Tsai (Eds.): Optical Nanotechnologies,
Topics Appl. Phys. **88**, 87–108 (2003)

However, these techniques use far-field light passing through the small aperture.

Another attractive method to increase optical data-storage density is to utilize optical near-field recording. This technology has the potential to increase storage density, even up to the terabyte level. Optical near-field recording was first proposed by *Betzig* et al. [20]. In their work, a near field is formed from a small aperture opened up at the tip of the metal-coated fiber. The distance between the aperture and the surface is less than 10 nm, which is controlled by the shear force. Thus, the recorded mark size is normally determined by the aperture size. Until now, much effort has been put into producing a smaller aperture and increasing the transmittance [21,22]. Nevertheless, the data transfer rate at present is too slow to apply the technique to real data-storage devices. On the other hand, space control by the shear force is not applicable, especially for rotating a disk at a high speed as for compact disks (CD) or DVDs. The main obstacle is the tip crush. In order to apply near-field optical recording to practical applications, *Terris* et al. [23] developed a technology that employs a solid immersion lens (SIL) to decrease the mechanical damage caused by the near-field optical fiber probe, and to achieve higher recording speed. However, in the past few years, the control of the near-field distance between the SIL recording head and the recording medium, contamination, and the near-field aperture size of the SIL have been the major hurdles for commercial applications.

Recently, *Tominaga* et al. proposed the new approach of the super-resolution near-field structure [24,25,26,27]. They named this structure super-RENS. In such a structure, a mask layer of Sb is deposited very close to the phase-change recording layer within the range of the near field. During the recording and readout processes, a small aperture that functions as a local near-field SIL is formed in the mask layer. Since the near-field distance is well controlled by a spacing layer between the mask layer and the recording layer, unlike the previous near-field storage techniques, the difficulties of the feedback control of the near-field distance and the degradation of the near-field probe can be overcome. Super-RENS is considered a more feasible way of near-field optical recording with simpler recording head design, less mechanical damage, no contamination and higher recording speed.

More recently, *Tominaga* et al. [28,29] demonstrated that a AgO_x-type super-RENS with an AgO_x layer as a mask layer has a much stronger near-field intensity and better carrier-to-noise ratio (CNR). The metallic probe was produced in an AgO_x mask layer. Small marks of less than 100 nm in length could be recorded and reproduced at a linear velocity of 6 m/s by the optimized AgO_x layer. However, the overwriting cycle is a big challenge for this type of super-RENS.

So far, the working mechanism of the super-RENS has not been well understood. *Fukaya* et al. [30] studied the optical switching property of a light-induced pinhole in the Sb thin film. *Tsai* et al. and *Liu* et al. investigated the focused spot size of Sb and AgO_x for understanding the working mechanism of the super-RENS in the transmission mode using a tapping-mode near-field scanning optical microscope (NSOM) [31,32,33,34,35,36]. The direct near-field optical imaging provided important information involved in super-RENS. *Tsai* et al. proposed the local surface plasmon (LSP) model to explain the working mechanism of the super-RENS [34].

Recent research has focused on searching for new mask materials, such as Zn and Sb_2Te_3, and new structures [37,38,39], understanding the working principle, improving the recording performance of the super-RENS such as the CNR and thermal stability.

2 Description of the Reading and Recording Process for the Super-RENS and the Requirements for the Mask-Layer Material and the Super-RENS

There are two main types of super-RENS. One is the Sb type (or melting type) and another is the AgO_x type (or scattering type). Figure 1 shows the Sb super-RENS disk structure.

2.1 Description of the Reading and Recording Process for the Super-RENS

For the Sb-type super-RENS, the film changes from the crystalline state to the liquid state when the temperature of the mask layer rises above the melting point during the recording process. A small hole is then formed. The hole can be considered as a small lens. Since the reflectivity of the material in the liquid state is much lower than that of the solid crystalline state, the rear portion of the laser spot is masked by the mask layer.

Tominaga et al. experimentally confirmed the formation of a temporarily light-induced pinhole in an Sb film by irradiation with a 10 ns pulsed laser beam. Pinhole formation seemed to be very rapid and started within the duration of the laser pulse and probably on a subnanosecond scale. With the

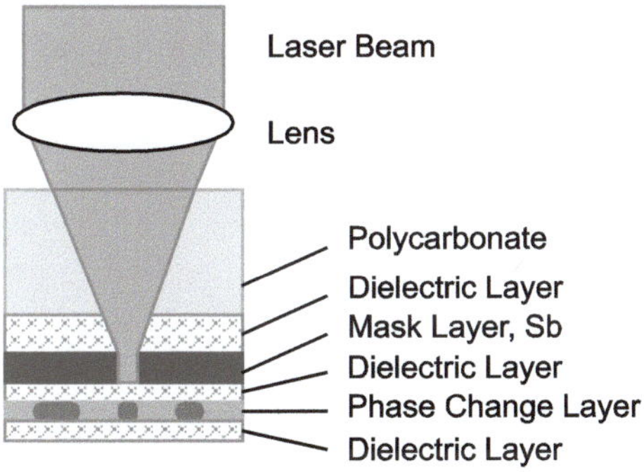

Fig. 1. Sb-type super-RENS disk structure

increase in input light intensity the transmittance of the Sb film changed drastically. When a laser beam was focused on the Sb film, a pinhole with a diameter smaller than that of the laser-beam spot was formed temporarily. As a result, optical near-field recording and readout through this pinhole have been confirmed.

Figure 2 shows the AgO_x super-RENS disk structure. In this structure, the AgO_x layer is used to produce the metallic probe. The AgO_x rapidly decomposes into Ag and oxygen in a small area heated above the threshold temperature by a focused laser beam. After the laser beam is removed, the Ag and oxygen form the AgO_x compound again. The decomposed Ag aggregates and produces a metallic area. The Ag area is not transparent and can be seen as a scattering center. Therefore, the near-field light is generated around it. This near-field light interacts with a recorded mark in the recording layer and scatters there, because the distance between the readout layer and the recording layer is within the range of the near-field light. Then, part of the scattered light is reflected back to the pickup. Thus, small marks can be recorded and reproduced through the near-field light around the Ag area. Hence, the decomposed Ag functions as a metallic probe for near-field recording and readout. On the other hand, part of the incident far-field light passes through the recording and readout layers because of low reflectivity, and the rest of the incident far-field light is absorbed to increase the readout layer temperature and is reflected back to the pickup in order to focus and track the laser beam.

It should be pointed out that from the chemistry point of view if the reaction speed is faster in one direction, the reaction speed in the opposite direction will be slower. This means that if the AgO_x decomposition speed is faster, the re-composition from Ag and O to form AgO_x will be slower. This is contrary to the experimental observations. Until now this phenomenon has not been well understood.

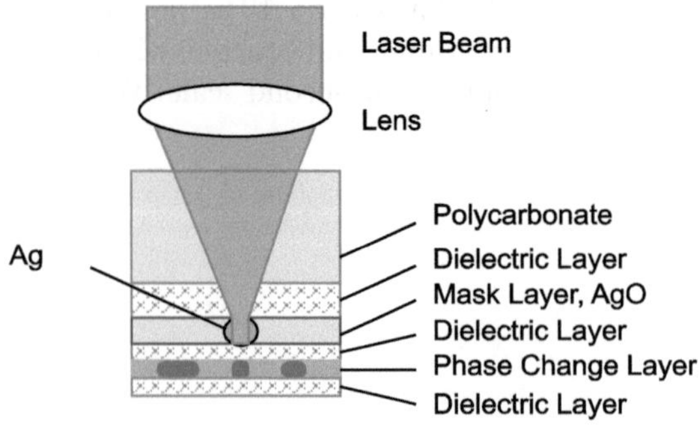

Fig. 2. AgO_x super-RENS disk structure

2.2 Requirements for the Mask Layer

In order to realize the super-RENS effect, several requirements for the mask layers must be met. Since a mask layer is used not only for recording but also for reading, it must have a fast reaction speed. This fast reaction is required for both the leading part and trailing part. If the leading part reaction is not fast, the distribution of the laser-beam intensity profile will be affected and in turn the recording and reading quality. Trailing-part reaction speed also affects the quality of the recording and reading signal. Compared to the trailing part, the leading-part reaction speed is more important.

Thermal stability is a critical issue for the super-RENS because the distance between the mask layer and the recording layer is very short. The change in the mask layer should not influence the state of the recording layer.

Compared to the conventional phase-change optical disk, the reading power for the super-RENS is higher. Therefore, the overwriting and over-reading cycle become important issues.

3 New Mask Material

The resolution of the super-RENS is very sensitive to the dielectric layer sandwiched between the mask and phase-change layers. With a Sb mask layer, the dielectric layer of SiN can achieve marks of 60 nm dimension, while ZnS–SiO_2 cannot. GeSbTe is conventionally used as a recording material and ZnS–SiO_2 is used as a dielectric material for phase-change disks. The invention of ZnS–SiO_2 as a dielectric layer was one of the most significant factors for achieving cycling times of millions of cycles [40].

It is well known that the GeSbTe system can be seen as pseudobinary GeTe and Sb_2Te_3 alloys with different combinations, such as $Ge_2Sb_2Te_5 \leftrightarrow (GeTe)_2(Sb_2Te_3)$, $Ge_1Sb_2Te_4 \leftrightarrow (GeTe)(Sb_2Te_3)$, and $Ge_1Sb_4Te_7 \leftrightarrow (GeTe)(Sb_2Te_3)_2$ (Fig. 3). This fact suggests to us the use of Sb_2Te_3 as a mask layer. Sb_2Te_3 has a rhombohedral lattice of the tetradymite (Bi_2Te_2S) type

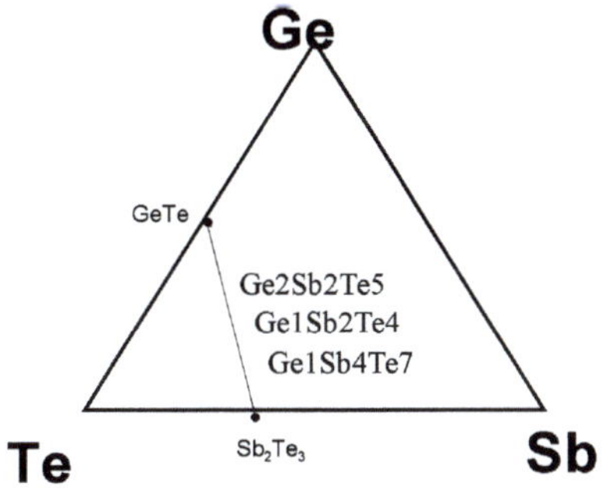

Fig. 3. Formation diagram of the GeSbTe system

(space group $R3m$) in the hexagonal configuration with the lattice parameters: $a = 0.4264$ nm and $c = 3.0453$ nm [41]. The hexagonal unit cell contains three five-layer packs ($N = 15$). The atomic layers are alternated in the $Te_1SbTe_2SbTe_1$ sequence perpendicular to the c axis. The five-layer packets are bonded to each other by weak van der Waals forces. The other promising characteristics of Sb_2Te_3 include its higher crystallization speed and lower crystallization temperature compared with those of Sb material [42,43]. Figure 4 shows the differential scanning calorimetry (DSC) curve of as-deposited Sb_2Te_3. From Fig. 4 it can be seen that as-deposited Sb_2Te_3 is in the crystalline state.

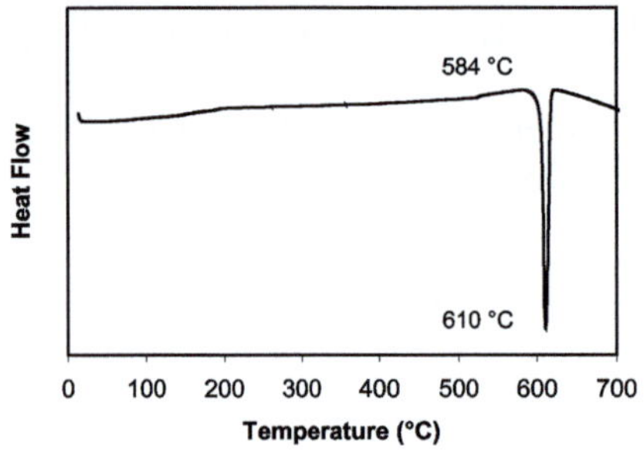

Fig. 4. DSC curve of as-deposited Sb_2Te_3

4 New Structures

Figures 5 and 6 show two new structures of the super-RENS. Figure 5 shows the plasmon-coupling layer between the mask layer and the recording layer, and Fig. 6 shows the plasmon-coupling layer in front of the mask layer. GeSbTe was used as the phase-change material and Sb_2Te_3 as the mask-layer material. ZnS–SiO_2 was used as dielectric material. For the plasmon-coupling layer, a highly conductive material was used. The main purpose of the new additional layers is to improve the thermal stability of the super-RENS. Moreover, it can also be used to induce localized surface-plasmon coupling.

5 Theoretical Simulation

Optical simulation was performed to evaluate the effect of the super-RENS. The modulation transfer function (MTF) can be derived from scanning microscope theory [44,45].

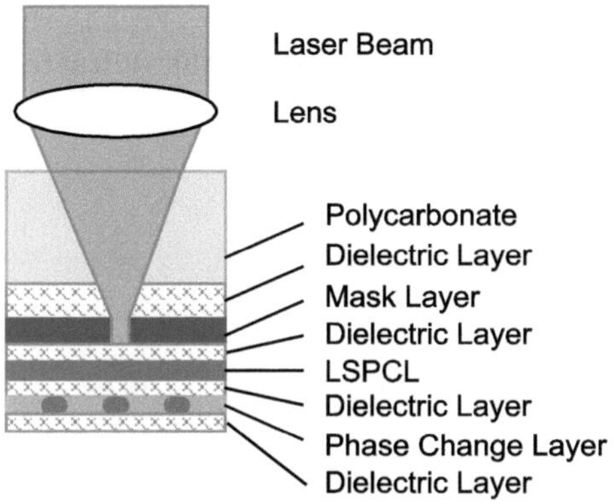

Fig. 5. New structure 1, with localized surface-plasmon coupling layer (LSPCL) between the mask layer and the recording layer

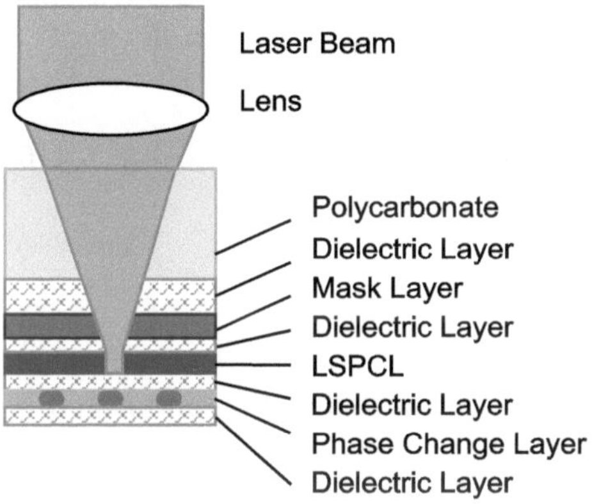

Fig. 6. New structure 2 with localized surface-plasmon coupling layer (LSPCL) in front of the mask layer

5.1 Optical Simulation

5.1.1 Modeling and Equations Used for Optical Simulation

The point-spread function is given by

$$h(u,v) = \int\int_{-\infty}^{\infty} P(x,y) \exp\left[2\pi \mathrm{i}(ux+vy)\right] \mathrm{d}x\mathrm{d}y\,, \tag{1}$$

where $P(x,y)$ is the pupil function of the objective lens and is given by

$$P(x,y) = \begin{cases} \exp\left[-\left(\frac{x^2}{W_x^2}+\frac{y^2}{W_y^2}\right)\right] & \text{if } (x+y)^2 \leq 1 \\ 0 & \text{elsewhere}\,. \end{cases} \tag{2}$$

W_x and W_y are the parameters that determine the amplitude distribution of the laser beam at the entrance aperture of the objective lens. Here W_x and W_y are assumed to be equal to each other and are chosen such that the intensity of the laser beam at the edge of the objective lens is 25% of that at the lens center. The amplitude distribution of the light leaving the disk can be given by

$$A_\mathrm{d}(u,v,u_\mathrm{s},v_\mathrm{s}) = h(u,v)\; f(u,v)\; t(u-u_\mathrm{s},v-v_\mathrm{s})\,, \tag{3}$$

where (u_s, v_s) is the displacement of the disk and $t(u-u_s, v-v_s)$ is the transmittance of the disk. The complex amplitude of the light diffracted backward to the point (x,y) on the pupil can be written as

$$\begin{aligned} A(x,y,u_s,v_s) &= \iint_{-\infty}^{\infty} h(u,v)\ f(u,v)\ t(u-u_s, v-v_s) \\ &\quad \times \exp[-2\pi \mathrm{i}(xu+yv)]\mathrm{d}u\mathrm{d}v \\ &= \iint_{-\infty}^{\infty} P(x-m, y-n) \otimes F(x-m, y-n) \times T(m,n) \\ &\quad \times \exp[-2\pi \mathrm{i}(u_s m + v_s n)]\ \mathrm{d}m\mathrm{d}n\,, \end{aligned} \tag{4}$$

Where the symbol $\otimes$ denotes the convolution operation and $T(m,n)$ and $F(m_f, n_f)$, which are the Fourier transform of t and f, are given by

$$T(m,n) = \iint_{-\infty}^{\infty} t(u,v) \exp[-2\pi \mathrm{i}(um+vn)]\ \mathrm{d}u\mathrm{d}v \tag{5}$$

and

$$F(m_f, n_f) = \iint_{-\infty}^{\infty} f(u,v) \exp[-2\pi \mathrm{i}(um_f + vn_f)]\ \mathrm{d}u\mathrm{d}v\,, \tag{6}$$

where m and m_f are the spatial frequencies in the x-direction and n and n_f are the spatial frequencies in the y-direction.

If introducing a new parameter, called the extended pupil function, the meaning of (4) becomes clearer to represent the convolution function in (4), i.e.,

$$P_{ex}(x-m, y-n) = P(x-m, y-n) \otimes F(x-m, y-n)\ . \tag{7}$$

The total light diffracted back to the pupil can be obtained as

$$\begin{aligned} I(u_s, v_s) = \iiiint_{-\infty}^{\infty} & C(m,n,m',n')\ T(m,n)\ T_m^*(m',n') \\ & \times \exp\{[-2\pi \mathrm{i}(m-m')u_s + (n-n')\,v_s]\}\ \mathrm{d}m\mathrm{d}n\mathrm{d}m'\mathrm{d}n'\,, \end{aligned} \tag{8}$$

where

$$\begin{aligned} C(m,n,m',n') = \iint_{\text{lens aperture}} & P_{ex}(x-m, y-n) \\ & P_{ex}^*(x-m', y-n')\ \mathrm{d}x\mathrm{d}y\,. \end{aligned} \tag{9}$$

The MTF in a given direction can be written as

$$\begin{aligned} \mathrm{MTF} &= |C(m,0,0,0)| \\ &= \left| \iint_{\text{lens aperture}} P_{ex}(x-m, y)\ P_{ex}^*(x,y)\ \mathrm{d}x\mathrm{d}y \right| . \end{aligned} \tag{10}$$

5.1.2 Optical Simulation Results

The simulation results show that the optical disk exhibits the maximum response when almost half the laser spot is covered by the mask. It can hardly be achieved with the conventional single-beam readout technique. However, it is possible to realize the faster response with the super-RENS by optimizing the disk structure and reading power to match a certain speed. This will be shown in the thermal simulation results.

5.2 Thermal Simulation

The cross-section of the super-RENS is shown in Fig. 1. In this structure there is a mask layer. Between the mask layer and the phase-change layer there is another dielectric layer. It can be seen that there are two heat-generation layers. Therefore, the thermal performance is quite different from that of conventional optical disks.

5.2.1 Modeling and Equations Used For Thermal Simulation

The fundamental equation based on linear transient thermal conduction is expressed as follows:

$$\nabla^2 T(r,t) + \frac{1}{k} g(r,t) = \frac{1}{\alpha} \frac{\partial T(r,t)}{\partial t}, \tag{11}$$

where $\alpha = k/\rho C_\mathrm{p}$ = thermal diffusivity, T is the temperature, k is the thermal conductivity coefficient. The term $g(r,t)$ gives the amount of heat energy per unit time and volume generated in the layer. The procedure for calculating $g(r,t)$ is expressed as follows. The laser beam intensity distribution is assumed to be Gaussian:

$$I = I_0 \exp\left(-\frac{r^2}{r_0^2}\right), \tag{12}$$

where I_0 is the incident light intensity at the center of the laser beam, $I_0 = P/(\pi r^2)$, P is the laser power and r_0 is the beam radius.

Taking into consideration the heating effects due to both the incident and reflected laser light, the term $g(r,t)$ can be divided into two terms written as follows (see Fig. 7 for the coordinates of the multilayer structure of the super-RENS):

$$\begin{aligned} I_1(r,t) &= A_1 I_0 \beta_\mathrm{mask}\, s(t) \exp\left(-\frac{(x-vt)^2+y^2}{r_0^2}\right) \\ &\quad \times \exp[-\beta_\mathrm{mask}(z-z_2)], \end{aligned} \tag{13}$$

$$\begin{aligned} I_2(r,t) &= A_2 I_0 \beta_\mathrm{mask}\, s(t) \exp\left(-\frac{(x-vt)^2+y^2}{r_0^2}\right) \\ &\quad \times \exp[-\beta_\mathrm{mask}(z_3-z)], \end{aligned} \tag{14}$$

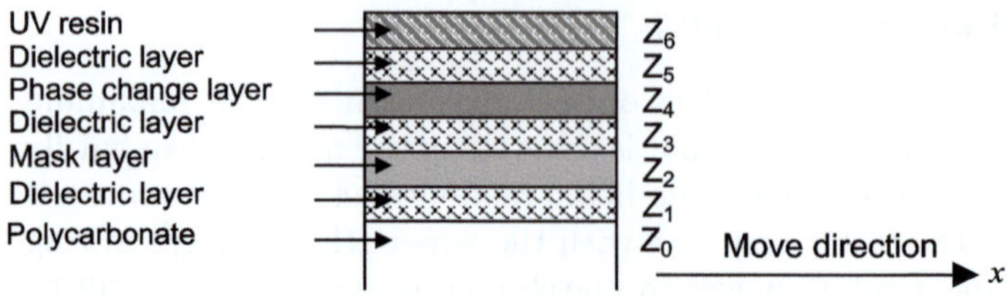

Fig. 7. Coordinates of the multilayer structure of the super-RENS

$$I_3(r,t) = A_3 I_0 \beta_{\mathrm{PC}}\, s(t) \exp\left(-\frac{(x-vt)^2+y^2}{r_0^2}\right) \exp[-\beta_{\mathrm{PC}}(z-z_4)]\,, \quad (15)$$

$$I_4(r,t) = A_4 I_0 \beta_{\mathrm{PC}}\, s(t) \exp\left(-\frac{(x-vt)^2+y^2}{r_0^2}\right) \exp[-\beta_{\mathrm{PC}}(z_5-z)]\,, \quad (16)$$

$$g_{\mathrm{mask}}(r,t) = \frac{I_1+I_2}{\pi r_0^2}\,, \quad (17)$$

$$g_{\mathrm{PC}}(r,t) = \frac{I_3+I_4}{\pi r_0^2}\,, \quad (18)$$

where I_1 and I_2 represent the contributions from the propagating and reflection light in the mask layer, respectively, and I_3 and I_4 represent the contributions from the propagating and reflection light in the phase-change layer, respectively. $A_1, A_2, A_3,$ and A_4 are the attenuation factors, which can be calculated by the optical method.

In the super-RENS structure, a mask layer and other dielectric layers are added. Therefore, both the mask layer and the phase-change layer will absorb laser energy. As a result the thermal performance is quite different from that of conventional optical disks. The thermal characteristics of the super-RENS structure can be described as follows:

1. There are two layers of heat sources. One is the mask layer and the other is the phase-change layer.
2. Before the mask layer reaches its melting point, the laser beam intensity distribution is Gaussian.
3. When the temperature of the mask layer reaches its melting point, an optical aperture is formed. The laser beam intensity distribution will then be changed. The focus beam can generate smaller marks in the phase-change layer.

5.2.2 Thermal Simulation Results

If the thickness of the dielectric layer sandwiched by the mask layer and phase change layers is too thin for the disks with the structures of Figs. 1 and 2, the temperature of the phase-change layer can be much higher than the crystallization temperature when the reading power heats the mask layer to the melting state. The simulation results show this behavior.

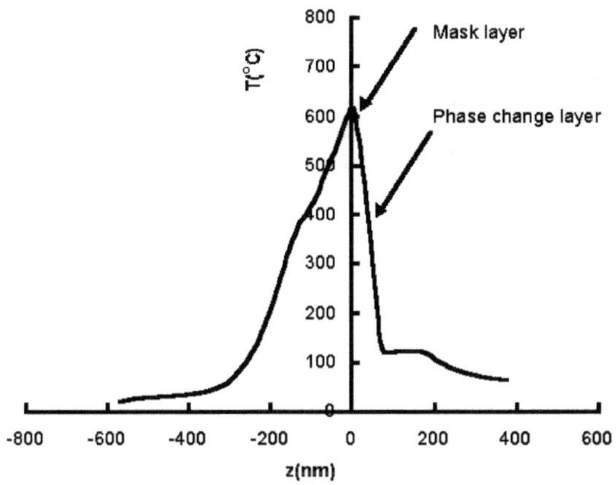

Fig. 8. Temperature distribution in new super-RENS structure 1

Temperature Distribution

Figure 8 shows the temperature distribution produced by the simulation. It can be seen that as the temperature reaches the melting point, the temperature in the phase-change layer becomes higher than the crystallization temperature. This means that the recorded marks can be erased. This simulation result has been proved by the experiments.

Figure 9 shows the temperature distribution in the new super-RENS structure 1. From these figures it can be seen that the temperature is much lower than that of the conventional super-RENS shown in Fig. 1.

Figure 10 shows the heat flow profile for the structures shown in Fig. 1 and Fig. 11 shows the heat flow profile for the new structure 2. In this simulation

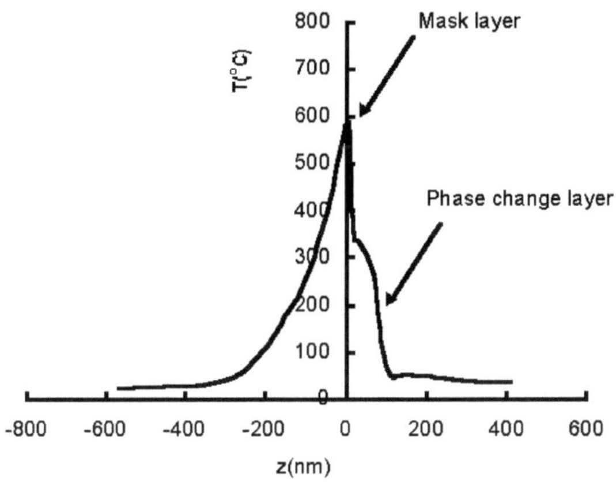

Fig. 9. Temperature distribution in a conventional super-RENS

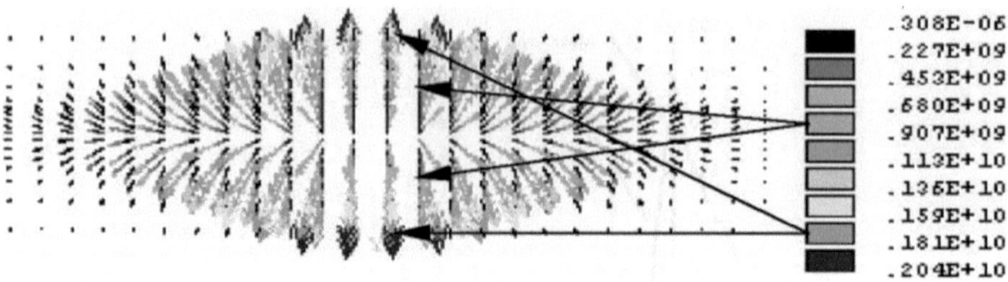

Fig. 10. Heat flow profile for structures shown in Fig. 1

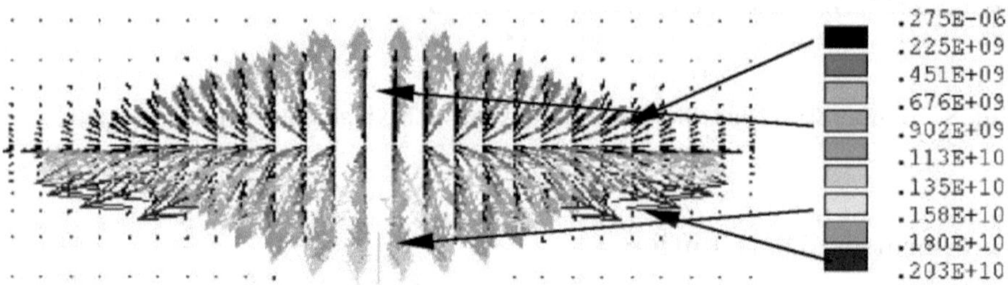

Fig. 11. Heat flow profile for new structure 2 shown in Fig. 6

it was assumed that the heat is only generated in the mask layer, so that we can clearly see the effect of the new thermal sink layer. Figure 10 shows that the heat flow rate is almost symmetrical in the conventional structure, whereas in Fig. 11 the heat generated in mask layer is mainly absorbed by the thermal sink layer. The main reason for this improvement is that the thermal sink layer is a good thermal conductor and the distance between the mask layer and the thermal sink layer is much smaller than that between the mask and the phase-change layer. In the reading process the heat will flow to the thermal sink layer. The simulation results demonstrated that the thinner is the thickness of the dielectric layer sandwiched by the mask layer, the more the heat flows to the thermal sink layer. The thicker is the thermal sink layer, the more the heat flows to the thermal sink layer.

Reading Optical Distribution at Different Conditions

Optical simulation shows that the optical disk exhibits the maximum response when almost half the laser spot is covered by the mask. This means that the super-RENS effect is related to the position between the laser beam and the hole melted in the mask layer. Figures 12, 13, 14, 15, 16 show the overlap of the contour line of the laser beam and the small hole. From the figures it can be seen that the overlap is related to the power and the speed of the disks. If the power and speed do not match, the hole will be very small, as shown in Fig. 16, or even cannot form. The higher is the power, the larger

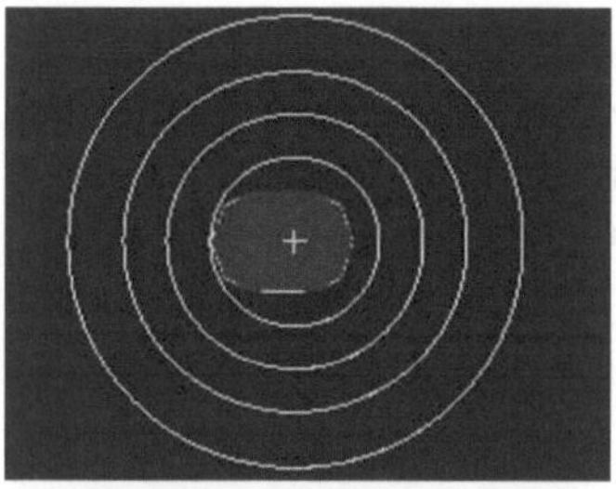

Fig. 12. Overlap of the contour line of the laser beam and the small hole at a reading power of 3.5 mW and velocity of 2 m/ s

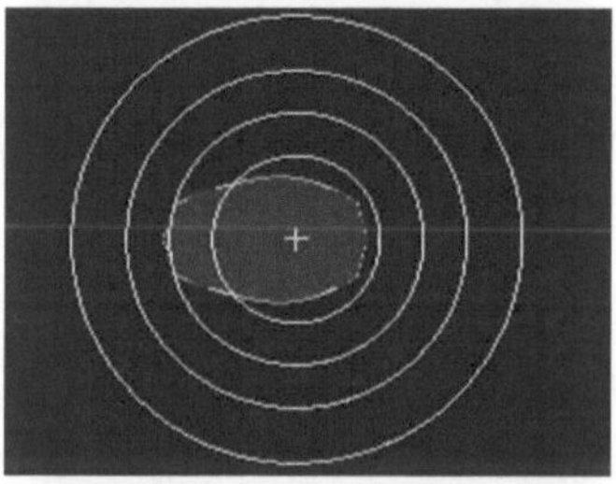

Fig. 13. Overlap of the contour line of the laser beam and the small hole at a reading power of 4.5 mW and velocity of 2 m/ s

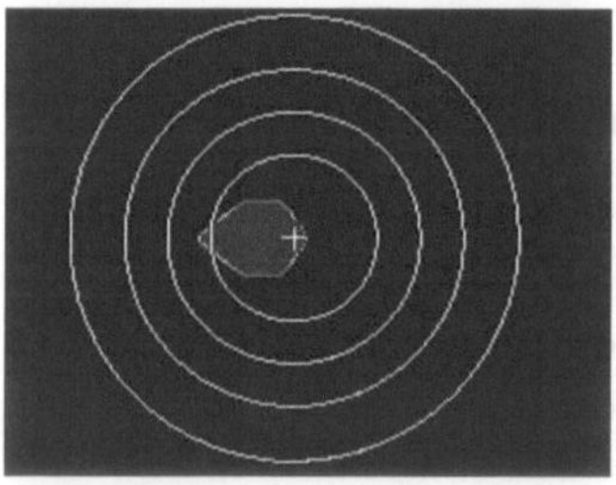

Fig. 14. Overlap of the contour line of the laser beam and the small hole at a reading power of 3.5 mW and velocity of 4 m/ s

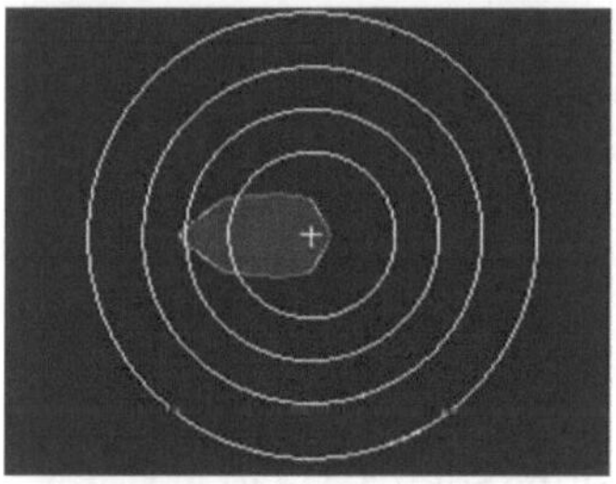

Fig. 15. Overlap of the contour line of the laser beam and the small hole at a reading power of 4.5 mW and velocity of 4 m/ s

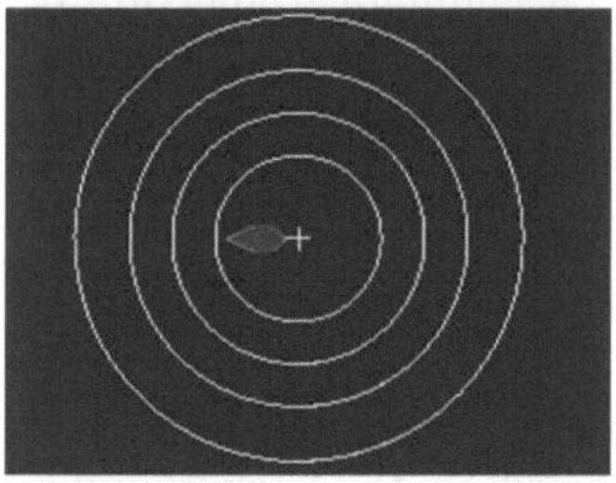

Fig. 16. Overlap of the contour line of the laser beam and the small hole at a reading power of 3.5 mW and velocity of 8 m/ s

is the hole. The faster the disk rotates, the smaller is the hole. On the other hand, this means that the overlap of the contour line of the laser beam and the small hole is also related to the disk structure and materials. It is clear that by optimizing the structure a faster response with the super-RENS can be achieved, as shown previously.

6 Fabrication of the Disk

The films were deposited on 0.6 mm thick polycarbonate substrates with a track pitch of 0.74 μm using the Balzer–Cube sputtering system. The PC recording layer, mask layer and plasmon-coupling layer were deposited by DC magnetron sputtering. The dielectric layers were deposited by RF sputtering. The background vacuum was at the level of 1.2×10^{-7} mbar. The working pressure was 4.5–5.5×10^{-3} mbar, with Ar as the processing gas at 15 sccm. After sputtering, the substrate with a multilayer stack was bonded to a blank substrate. The thickness was varied so as to observe the difference. Following three structures were fabricated:

1. Conventional structure:
   polycarbonate/ZnS–SiO_2(100 nm)/Sb_2Te_3 (10 nm)/ZnS–SiO_2(50 nm)/ GeSbTe (15 nm)/ZnS–SiO_2 (20 nm);

2. New structure 1:
   polycarbonate/ZnS–SiO_2(100 nm)/Sb_2Te_3 (10 nm)/ZnS–SiO_2(10 nm)/ plasmon-coupling layer (10 nm)/ZnS–SiO_2(40 nm)/GeSbTe (15 nm)/ ZnS–SiO_2 (20 nm);

3. New structure 2:
   polycarbonate/ZnS–SiO_2(100 nm)/plasmon-coupling layer (10 nm)/ ZnS–SiO_2(10 nm)/Sb_2Te_3 (10 nm)/ZnS–SiO_2(50 nm)/GeSbTe (10 nm)/ ZnS–SiO_2 (20 nm).

7 Measurement Results

The dynamic properties were measured using a ShibaSoku LM330 A tester. The wavelength of the laser beam was 650 nm. The numerical aperture of the objective lens was 0.6. The writing speed ranged from 0.6 to 6 m/ s, the reading power from 1 to 7.0 mW and the writing peak power from 8.0 to 18.0 mW.

7.1 Thermal Stability

Figure 17 shows the reflectivity versus time using a 2 mW reading power, where the thickness was 25 nm. In phase-change recording, the background is

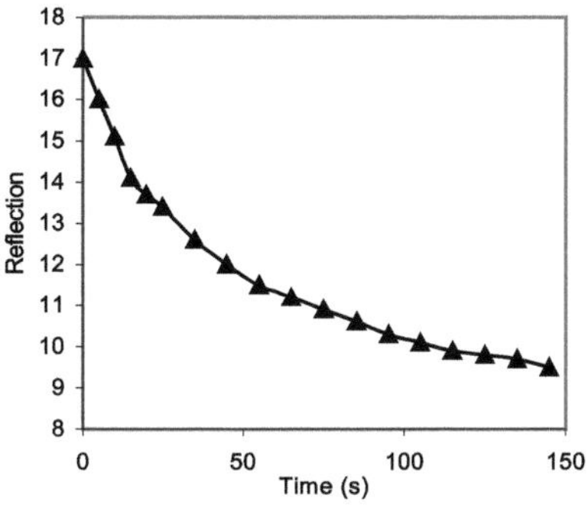

Fig. 17. Reflectivity versus time using a 2 mW reading power, where the thickness is 25 nm

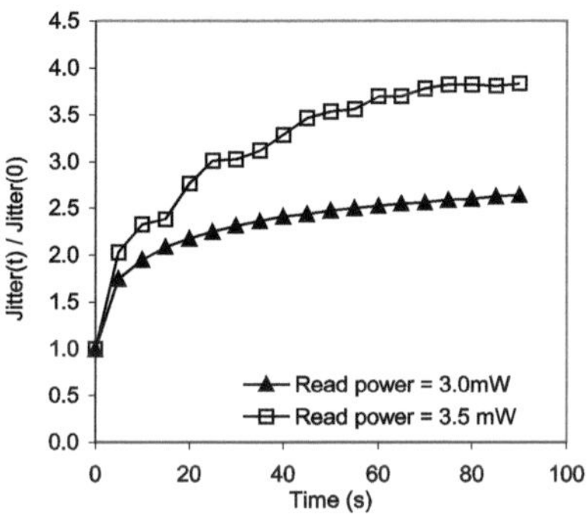

Fig. 18. Normalized jitter change with time for a disk with a 50 nm thick additional dielectric layer

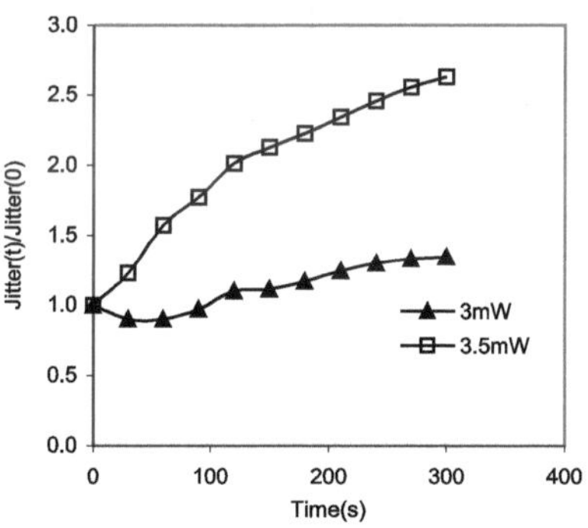

Fig. 19. Normalized jitter change with time for the new structure disk shown in Fig. 3 with a multilayer structure of 130/8/10/5/35/16/130 nm

in the crystalline state and the marks are in the amorphous state. The change of reflectivity reveals that the marks in the recording layer were erased in the reading process.

Figure 18 shows the normalized jitter change with time for the disk of Fig. 1 with a 50 nm additional dielectric layer sandwiched by the mask layer and the phase-change layer. Figure 19 shows the normalized jitter change with time for the new structure 1 with the multilayer structure of 130/8/10/5/35/16/130 nm. Comparing Fig. 18 with Fig. 19 it can clearly be seen that the thermal stability is significantly improved.

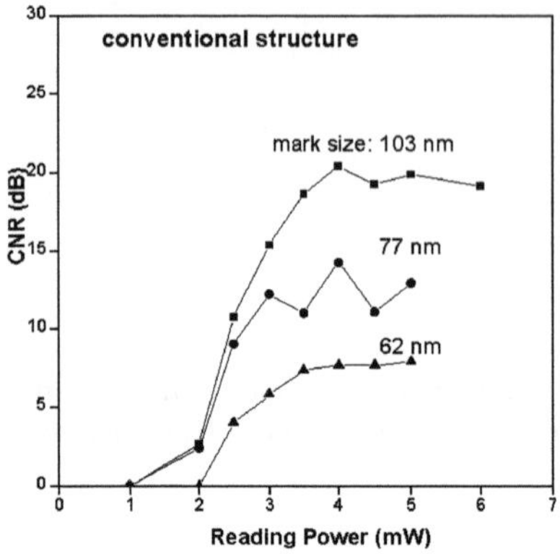

Fig. 20. CNR versus reading power at different mark lengths for the disks with the conventional super-RENS structure

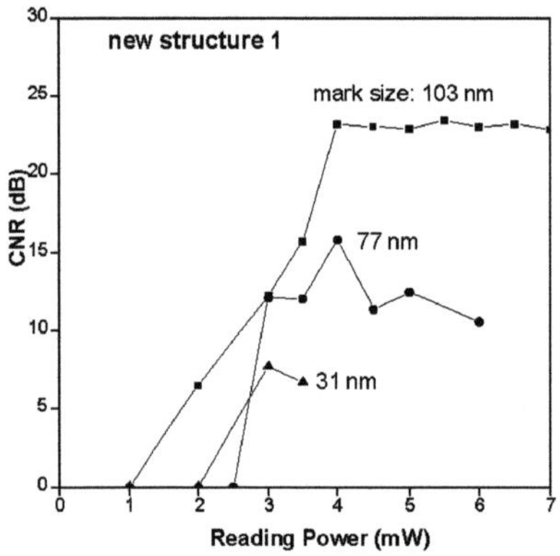

Fig. 21. CNR versus reading power at different mark lengths for the new structure 1 of the super-RENS disk

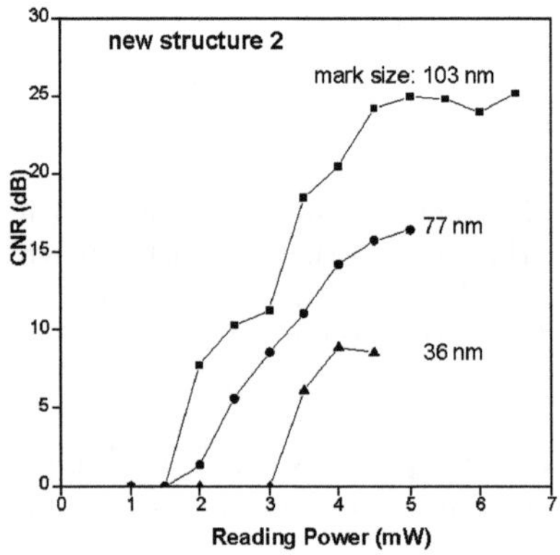

Fig. 22. CNR versus reading power at different mark lengths for the new structure 2 of the super-RENS disk

7.2 Super-RENS Effect

Figures 20, 21, 22 show CNR (Carrier Noise Ratio) versus reading power at different mark lengths for the disks with the above three structures. Figures 20, 21, 22 were obtained from the disks with the structures shown in Figs. 1, 5 and 6, respectively. For the conventional structure disk, the smallest mask size that can be observed is 56 nm and for the new structure disk, the smallest mask size that can be observed is 31 nm, which is roughly equal to the level of $\lambda/20$. The above results can be repeated, as each experiment was carried out at least three times including both disk fabrication and characterization.

It should be noted that the results were obtained by using red light. If we use blue light of 400 nm wavelength, the smallest achievable mark size would be expected to be about 20 nm.

8 Possible Mechanism of the Super-RENS

Until now, the most successful model to explain the working mechanism of the super-RENS has been the local surface-plasmon model of *Tsai* et al., although there are some issues that need to be further studied [34,35]. They used a tapping-mode tuning-fork near-field scanning optical microscope (TMTF-NSOM) to directly measure the images of the near-field intensity and the intensity gradients at different incident laser intensities in the Sb-type super-RENS. Their results showed that a focused laser could locally excite the super-RENS structure. The localized excitation area can be smaller than the laser-focused area. The area of the static evanescent intensity can be established in a very stable manner, and could be controlled by the focused laser power. The laser-excited area had large positive fluctuations locally, which occurred around the edge of the static evanescent intensity area. These results indicated that the formation of this evanescent intensity zone resulted from an ensemble of the localized surface plasmons, which were excited and controlled by the focused laser. The surface-plasmon enhanced local evanescent field acts as the near-field "aperture" described by *Tominaga* et al. [24,25,26,27]. The near-field recording and reading of the super-RENS are due to the local photothermal interactions between this excited evanescent area and the phase-change recording layer. The enhanced evanescent intensity may result from the localized surface plasmons excited at the mask and the dielectric layer interface. An ensemble of the localized surface plasmons and their photothermal energy transfer are believed to be the major factors of the super-RENS on the near-field optical storage.

Lie et al. [35] studied the transmission of the near-field intensity of a focused spot at the super-RENS sample, cover glass/ZnS–SiO_2 (20 nm)/AgO_x (15 nm)/ZnS–SiO_2 (20 nm), using a tapping-mode near-field scanning optical microscope.

The nonlinear optical properties and surface plasmon effect of the AgO_x-type super-RENS have been observed experimentally using a TMTF-NSOM. The TEM picture of the AgO_x-type super-RENS reveals that the AgO_x layer is heterogeneous. Numerical simulations indicate polarization dependence and strong localized peaks at the rough surfaces, which imply the existence of localized surface plasmons. The enhanced evanescent intensity may result from the localized surface plasmons excited at the interface. The complicated material structure further enhances the surface-plasmon effect and introduces nonlinear optical properties. The collective effects of the localized surface plasmons and the complicated material structure are suggested to be the working mechanism of the super-RENS for the near-field optical storage.

In both Sb- and Ag-type super-RENS, LSP plays an important role, yet only the mask layer functions as an LSP source to get small marks beyond the diffraction limit. If we introduce another LSP source into the conventional structure within the near-field distance, the two LSP sources may couple with each other. This coupled localized surface plasmon (CLSP) can be stronger than the LSP in the case of a single mask-layer, as shown in Fig. 1. As a result the smallest masks observed from the structures of Figs. 5 and 6 should be smaller than that observed from the structures of Fig. 1. This was confirmed by the much smaller mark observed in the new structures 1 and 2, when compared to the conventional structure. This CLSP may be the major reason for the observation of the small mark size.

It is notable that LSPs may also be formed in the crystalline state of the phase-change layer. Therefore, there are three LSP interactions for the structures in Figs. 5 and 6. The interactions of the three LSP sources and their photothermal energy transfer are related to the distance and the arrangement of the three LSP sources. This may result in the difference observed from Figs. 21 and 22. It should be pointed out that the interactions of the three LSPs are very complicated in the cases of Figs. 5 and 6. The mechanism of the coupling and control of the CLSPs are still important issues. The interactions of LSPs and their relationship with the materials and thickness of the mask layer, thermal-coupling layer, dielectric layer and recording layer need to be further studied. At least the above results provide indirect evidence to support Tsai's LSP model. Research on the small mark is based on this CLSP concept. Recently, *Tominaga* et al. [46] reported on the experimental observation of the local plasmon coupling effect between two mask layers in which there was a phase-change layer. A super-RENS disk with a second light-scattering center below the recording layer was designed. The second scattering center was also separated by inserting a ZnS–SiO_2 layer with a thickness in the range of the near-field interaction distance. The formation of the scattering centers in the two AgO_x layers was mainly due to heat propagation from the phase-change recording layer. Thus far, they have applied only a single AgO_x layer in LSC-super-RENS disks; however, by placing a second AgO_x layer on the opposite side, the layer may also generate a scattering center if the heat propagates uniquely in a narrow region of about 50 nm. In their experiment, the coupling phenomenon happened between three local surface plasmon sources, which is quite similar to the CLSP in the new structures 1 and 2.

Tsai et al. observed the large positive fluctuations around the edge of the static evanescent intensity zone. This may be caused by the imperfection of the localized excitation at the interface of the mask layer and the dielectric layer. The accuracy and perfection of the near-field recording marks induced by the super-RENS may be limited by this imperfect edge. The CNR of the super-RENS optical disk may also be significantly affected by the aperture edge problem. In order to investigate the imperfect edge, we measured the mask periphery by using TEM, as shown in Fig. 23. In some photos, the imperfect edge is similar to that shown in Fig. 23. This results is a direct observation of the imperfect edge that supports Tsai's indirect observation and his model. The imperfection of the edge of the excited spot or the evanescent intensity zone is one of the important limitations affecting the recording accuracy and density of the super-RENS. This mark edge irregularity is related to the grain size and the crystallization behavior. The imperfect edge in the mask layer can also be attributed to these two reasons. In fact, the large positive fluctuations around the edge of the static evanescent intensity zone observed by Tsai could be caused by the combination of the imperfect edges in the mask and phase-change layers.

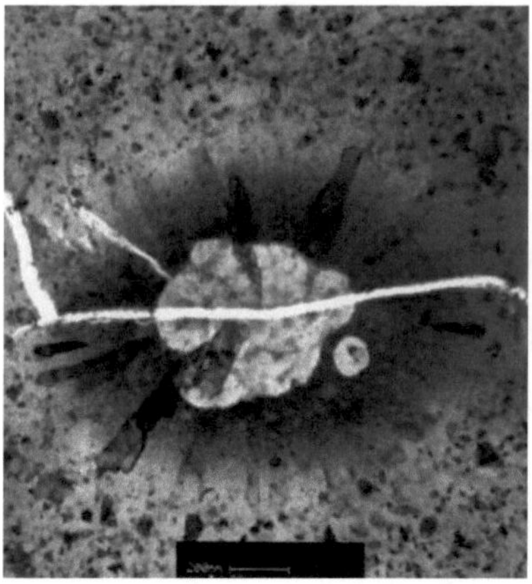

Fig. 23. TEM image of imperfect mark edge

9 Conclusions

The structure of a super-resolution near-field phase-change optical disk with localized surface plasmon coupling effect is proposed. A localized surface-plasmon coupling layer (LSPCL) was introduced to form a coupled localized surface plasmon (CLSP) with the mask layer. A new mask layer of Sb_2Te_3 was adopted. Recording marks as small as 31 and 36 nm were observed in two structures, which were both much smaller than the smallest mark of 56 nm observed in the conventional one without LSPCL. CLSP not only can reduce the mark size but it can also improve the carrier-to-noise ratio (CNR) of the recording marks. The local surface-plasmon coupling mechanism based on Tsai's LSP model was introduced to explain this phenomenon. The imperfect mark edge was observed by TEM. The thermal stability of the disk was also studied. Two new structures with a thermal shield layer that can significantly improve the thermal stability were proposed. The thermal stability of the new structures with a plasmon-coupling layer was much superior to that of the conventional structure.

References

1. N. Akahira, N. Miyagawa, K. Nishiuchi, Y. Sakaue, E. Ohno: Proc. SPIE **2514**, 294 (1995) 87
2. T. Ohta, M. Uchida, K. Yoshioka, K. Innoue, T. Akiyama, S. Furkawa, K. Kotera, S. Nakamura: Proc. SPIE **1078**, 27 (1989) 87
3. M. Horie, T. Ohno, N. Nobukuni, K. Kiyono, T. Hashizume, M. Mizuno: ODS 2001, Santa Fe, New Mexico, USA, 22–25 Apr. 2001, Tech. Dig. 37 (2001) 87
4. N. Yamada, E. Ohno, K. Nishiuchi, N. Akahira, M. Takao: J. Appl. Phys. **69**, 2849 (1991) 87
5. M.H. Kryder: J. Appl. Phys. **57**, 3913 (1985) 87
6. M. Mansuripur: *The Physical Principles of Magneto-Optical Recording* (Cambridge University Press, Cambridge 1995) 87
7. N. Ogihara, K. Shimazaki, Y. Yamada, M. Yoshihiro, A. Gotoh, H. Fujiwara, F. Kirino, N. Ohta: Jpn. J. Appl. Phys., Suppl. **28**, 61 (1989) 87
8. T.W. McDaniel, R.H. Victora: *Handbook of Magneto-Optical Data Recording* (Noyes, Westwood, NJ, USA 1997) 87
9. DVD specifications for rewritable Disc (DVD-RAM) version 1.0, DVD Format/Logo Licensing Corporation (July 1997) 87
10. M.J. Dekker, N. Pfeffer, M. Kuijper, I.P.D. Ubbens, W.M.J. Coene, E.R. Meinders, H.J. Borg: Proc. SPIE **4090**, 28 87
11. K. Nagata, N. Yamada, K. Nishiuchi, S.F.N. Akahira: Jpn. J. Appl. Phys. **38**, 1679 (1999) 87
12. K. Narumi, S. Furukawa, T. Nishihara, H. Kitaura, R. Kojima, K. Nishiuchi, N. Yamada: ISOM 2001, Taipei, Taiwan, 16–19 Oct. 2001, Tech. Dig., pp. 202 87
13. M. Kaneko, K. Aratani, M. Ohta: Jpn. J. Appl. Phys. **31**, 568 (1992) 87
14. S. Yoshimura, A. Fukumoto, M. Kaneko: Jpn. J. Appl. Phys. **31**, 576 (1992) 87
15. A. Takahashi, J. Nakajima, Y. Murakami, K. Ohta, T. Ishikawa: IEEE Trans. Magn. **30**, 232 (1994) 87
16. K. Yasuda, M. Ono, K. Aratani, A. Fukumoto, M. Kaneko: Jpn. J. Appl. Phys. **32**, 5210 (1993) 87
17. H. Awano, S. Ohnuki, H. Shirai, N. Ohta: Appl. Phys. Lett. **69**, 4257 (1996) 87
18. T. Shiratori, E. Fuji, Y. Miyaoka, Y. Hozumi: J. Magn. Soc. Jpn. **22**, 47 (1998) 87
19. T. Shintani, M. Terao, H. Yamamoto, T. Naito: Jpn. J. Appl. Phys. **38**, 1656 (1992) 87

20. E. Betzig, J. Trautman: Science **257**, 189 (1992); E. Betzig, J. Trautman, R. Wolfe: Appl. Phys. Lett. **61**, 142 (1992) 88
21. G. Valaskovic, M. Holton, G. Morrison: Appl. Opt. **34**, 1215 (1995) 88
22. N. Islam, X. Zhao, A. Said, S. Mickel, C. Vail: Appl. Phys. Lett. **71**, 2886 (1997) 88
23. B. D. Terris, H. J. Marnin, G. S. Kino: Appl. Phys. Lett. **65**, 388 (1994) 88
24. J. Tominaga, T. Nakano, N. Atoda: Appl. Phys. Lett. **73**, 2078 (1998) 88, 103
25. J. Tominaga, T. Nakano, N. Atoda: Proc. SPIE **3467**, 282 (1998) 88, 103
26. J. Tominaga, H. Fuji, A. Sato, T. Nakano, T. Fukaya, N. Atoda: Jpn. J. Appl. Phys. **37**, L1323 (1998) 88, 103
27. T. Nakano, A. Sato, H. Fuji, J. Tominaga, N. Atoda: Appl. Phys. Lett. **75**, 151 (1999) 88, 103
28. H. Fuji, J. Tominaga, T. Nakano, N. Atoda, H. Katayama: ISOM/ODS '99, Koloa, Hawaii, USA, July 1999, Tech. Dig., paper TuD29 88
29. H. Fuji, J. Tominaga, L. Men, T. Nakano, H. Katayama, N. Atoda: Jpn. J. Appl. Phys. **39**, 980 (2000) 88
30. T. Fukaya, J. Tominaga, T. Nakano, N. Atoda: Appl. Phys. Lett. **75**, 3114 (1999) 88
31. D. P. Tsai, S. Y. Lin, S. C. Yang, H. L. Huang, C. R. Chang, W. C. Lin, F. H. Ho, H. J. Huang, W. Y. Lin, W. C. Liu, T. F. Tseng, C. H. Li: ISOM 2000, Chitose, Hokkaido, Japan, Oct. 2000, Tech. Dig., paper We-D-02 89
32. D. P. Tsai, Y. Y. Lu: Appl. Phys. Lett. **73**, 2724 (1998) 89
33. D. P. Tsai, C. W. Yang, S. Z. Lo, H. E. Jackson: Appl. Phys. Lett. **75**, 1039 (1999) 89
34. D. P. Tsai, W. C. Lin: Appl. Phys. Lett. **77**, 1413 (2000) 89, 103
35. W. C. Liu, C. Y. Wen, K. H. Chen, W. C. Lin, D. P. Tsai: Appl. Phys. Lett. **78**, 685 (2001) 89, 103
36. D. P. Tsai, J. Kovacs, Z. Wang, M. Moskovits, J. S. Suh, R. Botet, V. M. Shalaev: Phys. Rev. Lett. **72**, 4149 (1994) 89
37. J. Park, H. Seo, T. H. Jeong: ODS 2000, Whistler, British Columbia, Canada, May 2000, Tech. Dig., postdeadline papers, pp. 29 89
38. L. P. Shi, T. C. Chong, X. S. Miao, P. K. Tan, J. M. Li: Jpn. J. Appl. Phys. **40**, 1649 (2001) 89
39. L. P. Shi, T. C. Chong, H. B. Yao, P. K. Tan, X. S. Miao: J. Appl. Phys. **91** (12), 10209 (2002) 89
40. N. Akahira, N. Miyagawa, K. Nishiuchi, Y. Sakaue, E. Ohno: Proc. SPIE **2514**, 294 (1995) 91
41. T. L. Anderson, H. B. Krause: Acta Cryst. B **30**, 1307 (1974) 92
42. X. S. Miao, T. C. Chong, L. P. Shi, P. K. Tan, F. Li: ISOM/ODS '99, Koloa, Hawaii, USA, July 1999, Tech. Dig., p. 285 92
43. N. Yamada, E. Ohno, K. Nishiuchi, N. Akahira, M. Takao: J. Appl. Phys. **69**, 2849 (1991) 92
44. Y. H. Wu, T. C. Chong: Appl. Opt. **36**, 6668 (1997) 92
45. T. Wilson, C. Sheppard: *Theory and Practice of Scanning Optical Microscopy* (Academic Press, London 1984) pp. 29–36 92
46. J. Tominaga, H. Fuji, D. Büchel, C. Mihalcea, T. Kikukawa, A. Sato, T. Nakano, A. Tachibana, M. Kumagai, A. Nomura, T. Fukaya, N. Atoda: ISOM 2001, Taipei, Taiwan, 22–25 Oct. 2001, Tech. Dig. pp. 40 104

Index

Polarization Dependence Analysis of Readout Signals of Disks with Small Pits Beyond the Resolution Limit

Takashi Nakano[1], Hisako Fukuda[1],
Junji Tominaga[1], and Takashi Kikukawa[2]

[1] Laboratory for Advanced Optical Technology (LAOTECH), National Institute of Advanced Industrial Science and Technology (AIST),
1-1-1 Higashi, Tsukuba, Ibaraki, 305-8562, Japan
t-nakano@aist.go.jp
[2] Information Technology Research Center, TDK Corp.,
462-1 Otai, Saku, Nagano 385-0009, Japan

Abstract. We describe the optical properties of the readout signal from light-scattering-center super-resolution near-field structure disks and super-resolution ROM disks by a numerical simulation using a three-dimensional finite-difference time-domain method. The calculated signals show a large polarization dependence. These results suggest the readout principle of small pits strongly related to the near-field optical phenomena including the local plasmons. We propose new optical pickup system designs using polarization dependence for the development of these high-density disks.

1 Introduction

Near-field optics has been applied to optical data storage achieving high recording densities beyond the diffraction limit. In this case, small apertures, scattering points or a solid immersion lens (SIL) have been used for recording or retrieving small marks beyond the diffraction limit. *Betzig* et al. first used a small aperture to record and retrieve small marks on magneto-optical media [1]. *Martin* et al. adapted an oscillating aperture-less probe for high frequency operation for high-speed readout [2]. Planar aperture flying heads have also been developed [3,4,5]. *Terris* et al. also applied SIL for magneto-optical recording [6]. *Milster* et al. analyzed SIL systems using numerical simulation and static experiments [7]. However, all the above systems have many problems. The difficulty of applying near-field optics to optical data storage is caused by the narrow space ($< 100\,\mathrm{nm}$) between the near-field probe and the recording medium.

In order to overcome this issue, we proposed another near-field optical technique called the "super-resolution near-field structure" (super-RENS). In this technique, a near-field probe similar to a small aperture or a light scattering center is generated in a mask layer, which is placed close to the recording layer [8,9]. We have succeeded in retrieving small marks beyond

J. Tominaga and D. P. Tsai (Eds.): Optical Nanotechnologies,
Topics Appl. Phys. **88**, 109–119 (2003)

the diffraction limit by a conventional optical disk drive tester. In super-RENS disks, the transmitted signals can also be detected at the opposite disk side [10]. This is strong evidence of near-field scattering in super-RENS disks, and the difference between the super-RENS disks and the conventional optical disks with a far-field super-resolution.

On the other hand, the super-resolution ROM (super-ROM) disk shows high-resolution characteristics with the enhancement of the signal intensity of small pits beyond the resolution limit with increasing readout power [11]. This principle has not been cleared but it is supposed that the characteristics are related to the near-field scattering in the nanometer region.

In the super-RENS and super-ROM disks, it is possible to readout small pits beyond the resolution limit using conventional optical pickups. However, the optical properties of the readout signals are different from those of a conventional one because the signal readout process is due to the interaction of a near-field probe or local plasmons generated in the disks with the pits. Therefore, it is supposed that the optical pickup systems are not optimized for these new disks to get good signals. In near-field optics, it is well known that many phenomena show polarization dependence. Therefore, the investigation of the polarization dependence of the readout signals of the super-RENS disks and super-ROM disks is important to improve the signal readout properties and to design optimum optical pickups.

In this paper, we describe the polarization dependence of the readout signals of super-RENS disks and super-ROM disks by numerical simulations using a finite-difference time-domain method (FDTD method).

2 Polarization Dependence of Readout Signals of Super-RENS Disks

We have already published the experimental results for the readout properties of super-RENS disks [12]. When we measured the transmitted signal from the light-scattering mode (LSC) super-RENS disks, the signal level of small marks beyond the resolution limit did not depend on the size of the numerical aperture of the pickup lens. This observed property is not explained by the theory of conventional optics. However, these experimental results included unknown properties of the recording process.

Therefore, the optical simulation of the near-field optics is very important to discuss the optical properties of the super-RENS disk [13]. Since conventional optical simulation programs are not directly applicable to the analysis of near-field phenomena, we have developed a near-field optical simulator by a three-dimensional FDTD method using the dispersion equation of a three-dimensional Lorentz model [14]. This FDTD method is a very useful tool for estimating the electromagnetic field and near-field interaction [15,16].

2.1 Simulation Model of a Super-RENS Disk

The LSC-super-RENS disk was composed of a ZnS–SiO_2/AgOx/ZnS–SiO_2/ $Ge_2Sb_2Te_5$/ZnS–SiO_2 multilayer on a substrate. A silver particle as a scattering center was generated in an AgOx layer by a focused laser beam under high-speed rotation. The super-RENS disks are expected to achieve a high carrier-to-noise ratio (CNR) by the local plasmon effect of the silver particle similar to a metallic probe near-field scanning microscope. In our simulation, we simulated the optical readout process after generating a silver particle. Therefore, the recorded marks and the silver particle were fixed in each layer. Figure 1 shows the simulation model of a LSC-super-RENS disk. This model is composed of a polycarbonate substrate /ZnS–SiO_2 (170 nm)/AgOx (15 nm)/ZnS–SiO_2(40 nm)/$Ge_2Sb_2Te_5$ (15 nm)/ZnS–SiO_2(20 nm) multilayer structure on a substrate. The thicknesses of the layers have the same values as those of an actual super-RENS disk. The refractive indices used in this calculation were measured by an ellipsometer at a wavelength of 633 nm. The substrate had a land and groove structure with a 600 nm width and 60 nm depth toward the x-direction.

The recorded mark size was fixed at 160 nm. The mark train with a 50% duty ratio exists at the center of the land. The size of a silver scattering center was fixed at 160 nm. The position of the scattering center was selected for two cases. One position is the center of the optical axis of the incident laser beam with a Gaussian distribution (coaxial model). Another position is shifted by 160 nm toward the tangential direction (x-direction) of the disk (shifted model), because we found that the scattering center and laser position are slightly different according to the thermal simulation.

A linearly polarized laser beam with a 650 nm wavelength is assumed as the incident light. The numerical aperture (NA) of the conversion lens was 0.6. In our 3D-FDTD calculation, we divided the model into $110 \times 110 \times 113$ unit cells. The size of each cell was $20 \times 20 \times 3\,\mathrm{nm}^3$ and the total volume was $2.2 \times 2.2 \times 0.34\,\mu\mathrm{m}^3$. The process time of one calculation was about three hours using a personal computer with an alpha processor (600 MHz, Digital

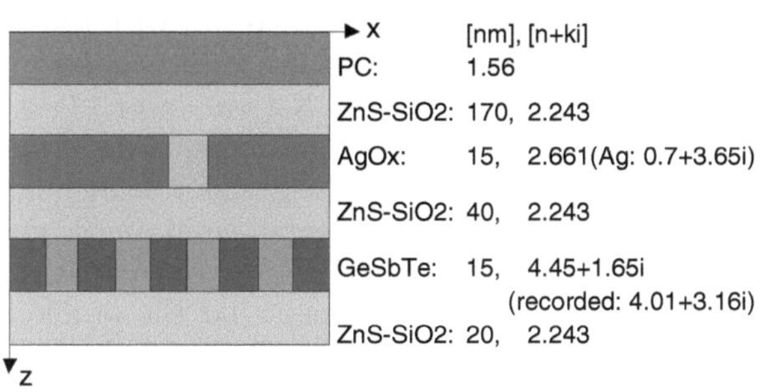

Fig. 1. Simulation model of a LSC-super-RENS disk

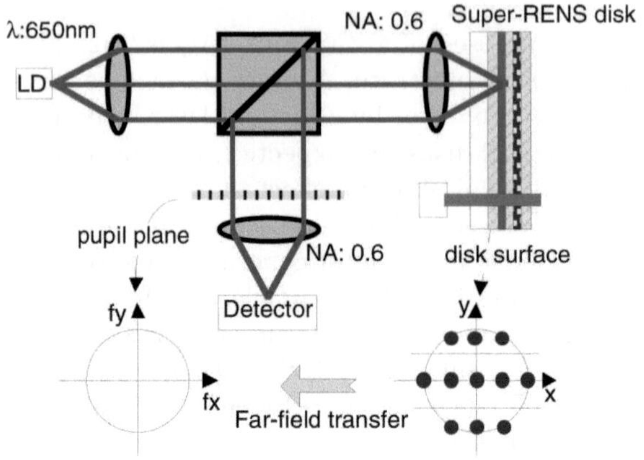

Fig. 2. Simulation model for reflected-light detection system

Equipment Corp.) and 1.5 GB memory. To estimate the readout modulation, electric fields at different mark positions were calculated.

After the 3D-FDTD calculation of the electromagnetic field, the propagated light intensity was estimated by a far-field transformation using a Fraunhofer approximation. Figure 2 shows the simulation model for the reflected-light detection system. In this model, the intensity distribution at the exit pupil plane of the objective lens was calculated to evaluate the readout signals. The exit pupil planes mean the Fourier planes of a focused plane (objective plane). A circle around the optical axis indicates the NA range of the observed objective lens.

2.2 Intensity and Signal Distribution at the Exit Pupil Plane

Figure 3 shows the intensity distributions in the exit pupil plane on the reflection side for a coaxial model. The silver particle is fixed just above one mark (on-mark). Figure 3a shows the result with the incident polarization along the mark trains (x-direction), and Fig. 3b shows the result with the incident polarization normal to the mark trains (y-direction). Three concentric circles indicate each exit pupil size determined by the NA values of 1.0, 0.6 and 0.3, respectively. The intensity profile is oriented toward the y-direction. In Fig. 3a, a strong peak area is clearly simulated in a small solid angle (NA = 0.3). In Fig. 3b, a principal peak exists in a small solid angle, and two sub-peaks also appear at the edges of the elliptical region. In comparison with a model without the land and groove, it is assumed that the profile is caused by the diffraction due to the land and groove. The intensity patterns hardly change by mark's shifts. This result indicates that the signal of the LSC-super-RENS disk is buried in a large background noise.

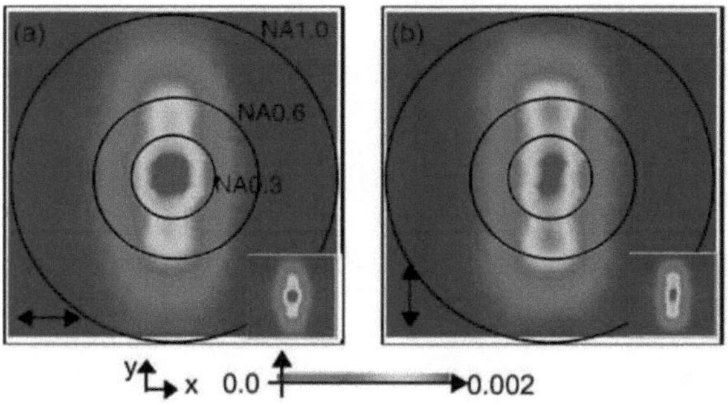

Fig. 3. Calculated intensity distribution in the pupil plane for a shifted scattering-center model by (**a**) x-polarized light and (**b**) y-polarized light

On the other hand, the intensity profile in the exit pupil plane of the shifted model did not show a large change in comparison with the previous results (coaxial model).

In order to extract the near-field signal from the background, the differences between the intensities of the on-mark and the off-mark cases were calculated. The off-mark means when the silver particle exists between marks. The results of the coaxial model are shown in Fig. 4. Figure 4a shows the result with the incident polarization along the mark trains (x-direction), and Fig. 4b shows the result with the incident polarization normal to the mark trains (y-direction). The maximum signal value is about 2% of the peak intensity in the same pupil planes. In Fig. 4, a central peak in a small solid angle has a positive value and the other two peaks at an intersection point of the y-axis and a 0.6 NA circle are negative. It is not better to use a high-NA lens, because the integrated signal has an offset. As shown in Fig. 4a, the absolute value of the positive peak is larger than that of the negative ones. On the other hand, as shown in Fig. 4b, the absolute value of the negative

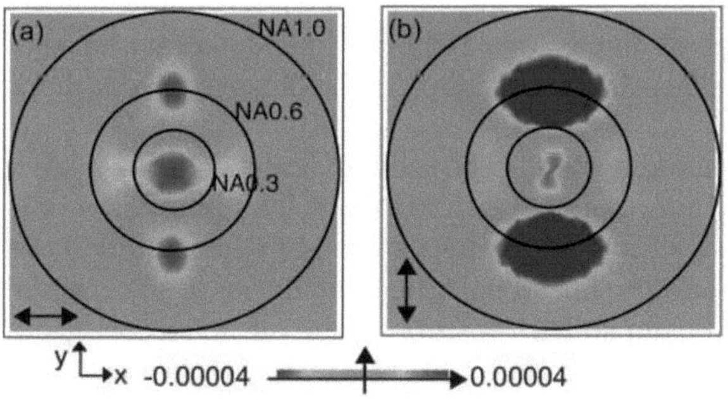

Fig. 4. Calculated signal distribution in the pupil plane for a coaxial model of a scattering-center position by (**a**) x-polarized light and (**b**) y-polarized light

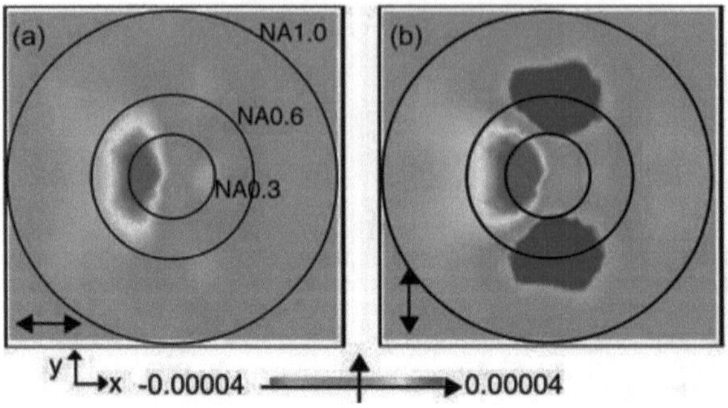

Fig. 5. Calculated signal distribution in the pupil plane for a shifted model of a scattering-center position by (**a**) x-polarized light and (**b**) y-polarized light

peaks is larger than that of the positive peaks. The distribution area of the negative peaks is larger than for any other case of incident polarization.

Figure 5 shows the calculated results of the distribution difference between the intensities of the on-mark and the off-mark cases of a shifted model. Figure 5a is the result with the incident polarization along the mark trains (x-direction) and (b) is the result with the incident polarization across the mark trains (y-direction). In Fig. 5a, a strong positive peak exists at the left edge of 0.3 NA. In Fig. 5b, one strong positive peak and two strong negative peaks along the y-direction also exist. The positive peak exists at the same position as that shown in Fig. 5a. In comparison with the previous results, the negative peak positions are not widely changed by the scattering center shift. On the other hand, the positive peak positions shift toward the x-direction.

The intensity and signal distribution pattern depend on the polarization direction of the incident light. The signal distribution pattern shows a scattering-center position dependence. However, the signal level is not changed by this shift.

2.3 Read-Out Signal Properties of a Super-RENS Disk

In order to evaluate the readout signal, we calculated the signal modulation by the integration of the intensity at the exit pupil plane of the different mark positions.

Figure 6a shows the calculated results of the coaxial model. The integrated region has an NA 0.6 (solid line) and 0.3 (dotted line). The horizontal axis shows the values of the mark shifts. The value of the shift is the difference between the shifted position and the initial position. The initial position is the on-mark condition and the 160 nm shifted position is the off-mark condition. The results shown in Fig. 6a indicate that the integrated intensity signals for different polarizations differ in phase by π. If the integrated region is reduced to the 0.3 NA region, the integrated signal for the y-polarization (square points) and that for the x-polarization (circular points) have the

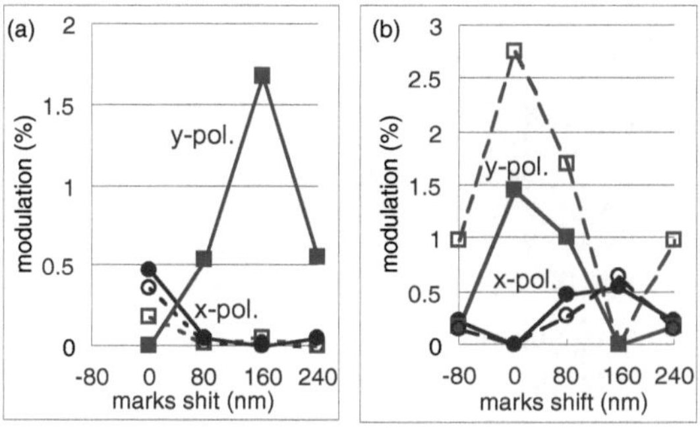

Fig. 6. Calculated signal modulation of (**a**) the coaxial model and (**b**) the shifted model of a scattering center position

same phase. This is because the integrated signal for the y-polarization is dominated by the two negative peaks shown in Fig. 4a,b. In this case, the result suggests that a polarization-discriminating method has the potential to further enhance the signal modulations.

Figure 6b shows the signal modulations calculated from Fig. 5a,b for a circular aperture (0.6 NA) by solid lines and for an annular aperture (NA: 0.3–0.6) by dotted lines. The modulation in the x-polarized incident-light case does not change in the detection aperture pattern. On the other hand, the modulation in the y-polarization case changes about twice. The annular detection pattern subtracts the positive peak from the integrated signals. The largest modulation is given by the annular detection of the y-polarized signals. It is five times larger than the minimum modulation by the x-polarization signal. The calculated signals for the different polarizations differ in phase by π with each other. When we calculate the modulation from the total intensity of the two polarization signals, the largest modulation is similar to the x-polarized one.

These results suggest that a polarization-discriminating method and optical filtering on the pupil plane have the potential to further enhance the signal modulation.

3 Polarization Dependence of the Read-Out Signals of the Super-ROM Disk

The conventional optical disk drive tester used circular polarized light for signal readout, except for the magneto-optical disks tester. However, we found very important properties of super-ROM disks using a linear polarized light readout. Figure 7 shows the carrier-to-noise ratio of 250 nm pit trains with a Ge reflective thin layer observed with a spectrum analyzer. This pit size is beyond the resolution limit of the experimental setup (VersaTest-1, NA: 0.6, wavelength: 660 nm). It is clear that the readout signals show the polarization

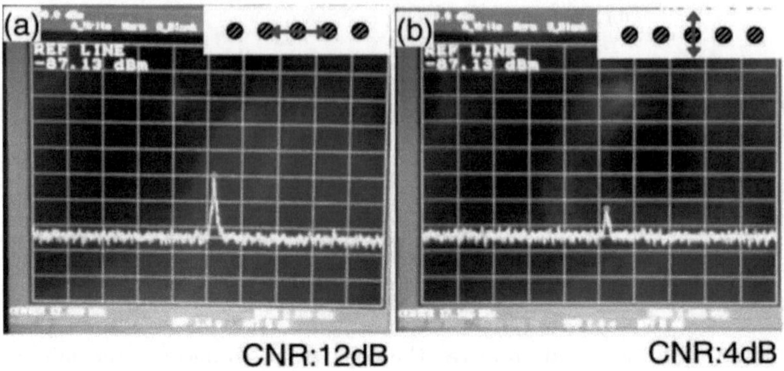

Fig. 7. Readout signal of 250 nm pits with a Ge reflector for (**a**) x-polarized light and (**b**) y-polarized light

dependence. The readout signal of x-polarized light parallel to the pit train (Fig. 7a, CNR: 12 dB), is higher than the signal readout of y-polarized light perpendicular to the pit train (Fig. 7b, CNR: 4 dB). This phenomenon was not observed at the disks with a Ag reflective layer.

These results suggest that the signal readout process is due to the interaction of a near-field probe or local plasmon generated in disks with small pits. To estimate this idea, we did the near-field optical simulations of signal readout.

3.1 Simulation Model of a Super-ROM Disk

Figure 8 shows the simulation model of super-ROM disks. This model consisted of a flat substrate and a thin silicon reflective layer with two pit patterns. One pit pattern consisted of 200 nm diameter pits with a 50% duty ratio (2T pattern). Another pattern consisted of a combination of 200 nm (2T) and 800 nm (8T) length pits and 200 nm space (H5 pattern). The pit width was fixed at 200 nm. The pit depth was 60 nm and the thickness of the reflection layer of silicon was 20 nm. The pit walls were coated with the reflection layer. The wavelength of the incident laser was fixed at 650 nm, the numerical aperture for the incidence and detection was assumed as 0.6. While changing the pit position, the electromagnetic field distribution was calculated. After the simulation, the intensity was integrated by a far-field transformation method under a Fraunhofer approximation.

3.2 Simulation Results of a Model with 2T Pit Pattern

Figure 9a and 9b shows the intensity distribution in the ROM disks with a Si reflective layer by x-polarized incident light and y-polarized incident light. Figure 9c and 9d shows the case of a Ag reflective layer. The square marks

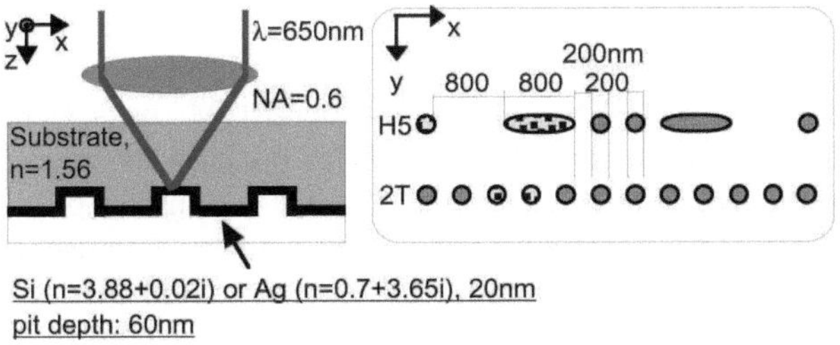

Fig. 8. Simulation model of a super-ROM disk and included pit pattern of 2T

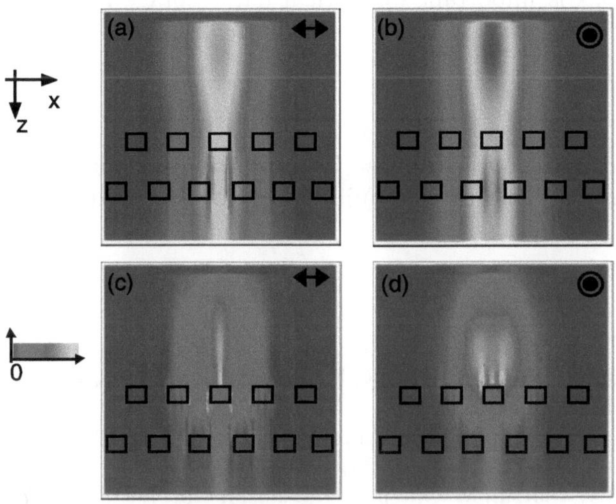

Fig. 9. Calculated intensity distribution in the ROM disk included 2T-pit pattern with Si reflector by linear polarized light of (**a**) x-direction and (**b**) y-direction, and with Ag reflector by linear polarized light of (**c**) x-direction and (**d**) y-direction

indicate the top and bottom of the pits (the scale of the x-direction is 7 times larger than the z-direction). In the case of the Si reflective layer, the incident light passes through the pit trains and the electric field of the pit walls is enhanced. This enhancement effect depends on the incident polarization of the light. On the other hand, in the case of the Ag reflective layer, the incident light is reflected at the top of the pit trains.

These simulation results show the same polarization dependence and material dependence as in the experimental results. The electric field enhancement in the pit trains is seen to have an important role for generating the small pit information. Therefore, it is important to apply the light along the x-polarization component for the readout pits.

4 Conclusion

The readout signals of LSC-super-RENS disks and super-ROM disks were evaluated numerical simulations. The calculated signals for both disks showed a large polarization dependence. In the super-RENS disks, annular pupil pattern detection is useful for obtaining high modulation signals. The use of a pickup with a polarizer and an annular pattern filter in the pupil plane is important for optical data storage in super-RENS disks. In the super-ROM disks, the difference intensity of the x- and y-polarization signals shows the real pit pattern information. The use of a pickup with two detectors for different light polarizations is important for optical data storage in super-ROM disks.

Acknowledgements

The authors thank Yuzo Yamakawa, Pioneer Co., for the development of this FDTD simulation program and Hiroshi Fuji, Sharp Co., and Akira Sato, Minolta Co. Ltd., for helpful discussions.

References

1. E. Betzig, J. Trautman, R. Wolfe, E. Gyorgy, P. Finn, M. Kryder, C. Chang: Appl. Phys. Lett. **61**, 142 (1992) 109
2. Y. Martin, S. Rishton, H. Wickramasinghe: Appl. Phys. Lett. **71**, 1 (1997) 109
3. F. Issiki, K. Ito, K. Etoh, S. Hosaka: Appl. Phys. Lett. **76**, 804 (2000) 109
4. T. Yatsui, M. Kourogi, K. Tsutsui, M. Ohtsu, J. Takahashi: Opt. Lett. **25**, 1279 (2000) 109
5. H. Yoshikawa, Y. Andoh, M. Yamamoto, K. Fukuzawa, T. Tamamura, T. Ohkubo: Opt. Lett. **25**, 67 (2000) 109
6. B. Terris, H. Mamin, D. Rugar: Appl. Phys. Lett. **65**, 388 (1994) 109
7. T. Milster, K. Shimura, J. Jo, K. Hirota: Opt. Lett. **24**, 605 (1999) 109
8. J. Tominaga, T. Nakano, N. Atoda: Appl. Phys. Lett. **73**, 2078 (1998) 109
9. H. Fuji, J. Tominaga, L. Men, T. Nakano, H. Katayama, N. Atoda: Jpn. J. Appl. Phys. **39**, 980 (2000) 109
10. T. Nakano, A. Sato, H. Fuji, J. Tominaga, N. Atoda: Appl. Phys. Lett. **75**, 151 (1999) 110
11. T. Kikukawa, T. Kato, H. Shingai, H. Utsunomiya: Jpn. J. Appl. Phys. **40**, 1624 (2001) 110
12. T. Nakano, T. Gibo, L. Men, H. Fuji, J. Tominaga, N. Atoda: Proc. SPIE **4085**, 201 (2001) 110
13. T. Nakano, Y. Yamakawa, H. Fuji, J. Tomianga, N. Atoda, Jpn. J. Appl. Phys. **40**, 1531 (2001) 110
14. J. Judkins, R. Ziolkowski: J. Opt. Soc. Am. A **12**, 1974 (1995) 110
15. L. Novotny, D. Pohl, R. Regli: J. Opt. Soc. Am. A **11**, 1768 (1994) 110
16. H. Furukawa, S. Kawata: Opt. Commun. **132**, 170 (1996) 110

Signal Power in the Angular Spectrum of AgOx SuperRENS Media

Tom Milster[1], John J. Butz[1], Takashi Nakano[2], Junji Tominaga[2], and Warren L. Bletscher[1]

[1] Optical Sciences Center, Optical Data Storage Center, University of Arizona, Tucson, AZ 85721, USA
milster@arizona.edu

[2] Laboratory for Advanced Optical Technology (LAOTECH), National Institute of Advanced Industrial Science and Technology (AIST), 1-1-1 Higashi, Tsukuba, 305-8562, Japan
t-nakano@aist.go.jp

Abstract. Light-scattering properties of super-RENS media are examined in the angular spectrum near the objective lens of an optical data storage device. Signal distributions resulting from mark patterns below the resolution limit, at the resolution limit and above the resolution limit are analyzed in order to understand better the mechanism of super-RENS resolution enhancement. Results from a combination of scalar and vector modeling tools are compared to measurements from a dynamic spin stand. Simulations assume that the super-RENS effect is generated from a circular Ag region that follows the focused spot as the medium moves underneath it.

1 Introduction

The super-resolution near-field structure (super-RENS) is an attractive technology that has good potential for improving surface density in optical data storage without the need for near-field probe structures [1,2,3]. Super-RENS technology effectively records and retrieves information beyond the resolution limit of classical optical systems by utilizing the properties of a nonlinear layer in close proximity to the data layer. The nonlinear layer responds to heat generated by a focused laser spot. The nonlinear layer can either open an aperture in an otherwise opaque film or produce a scattering center in an otherwise transparent film. The light-scattering-center type film is shown to exhibit stronger near-field intensity and better carrier-to-noise ratio. In this paper, we investigate the properties of the angular spectrum produced by the light-scattering-type nonlinear layer with both simulation and experiment. Our goal is to understand better the characteristics of the light-scattering center.

The basic light-scattering-center super-RENS system is shown in Fig. 1 for an AgOx nonlinear layer, which is transparent at room temperature. The complete structure includes three dielectric layers of ZnS–SiO_2, the AgO_x

J. Tominaga and D. P. Tsai (Eds.): Optical Nanotechnologies,
Topics Appl. Phys. **88**, 119–140 (2003)

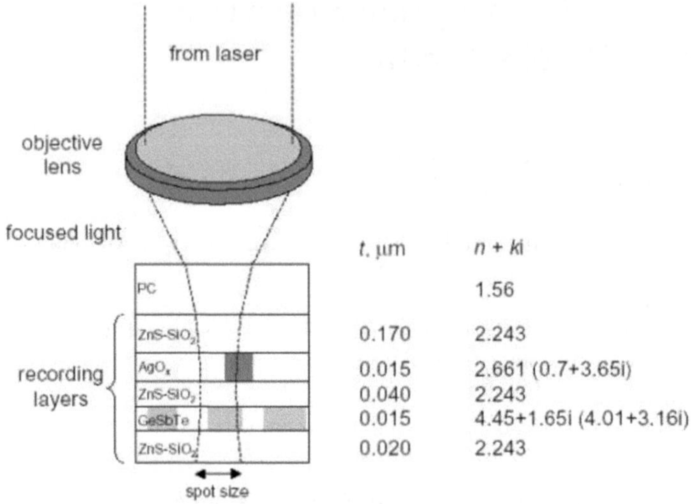

Fig. 1. Basic light-scattering-center super-RENS system and the structure of an AgOx super-RENS disk used in the simulations and experiments

layer and a layer of GeSbTe. The dielectric layers provide signal enhancement and environmental protection. The GeSbTe layer changes phase from crystalline to amorphous when a data mark is written into the material. Since the amorphous and crystalline phases have different optical reflectivity, the reflected laser light is modulated as the marks move under the focused spot. The recording layers are within the depth of focus of the focused spot.

As the disk spins, the spot heats a small area of the nonlinear layer to a threshold temperature, where Ag particles coalesce from the AgOx and form small light-scattering centers [4]. The region of Ag scattering centers is localized near the center of the spot, where the temperature reaches its maximum value. The scattering region may slightly trail the center of the light spot due to the conductive thermal properties of the material [5]. At some distance away from the center of the spot, the material cools, and the Ag particles return to their transparent state as AgOx. The presence of the scattering region acts to direct optical energy in a confined region at the GeSbTe layer. Energy confinement is significantly smaller than what is possible with far-field systems, thus enabling recording and readout of data bits well past the classical resolution limit [2].

The detailed nature of the interaction between the scattering region, the focused spot and the data pattern is not well understood. Several researchers are investigating different aspects of this problem for both aperture-type and light-scattering type super-RENS disks. For example, *Liu* et al. describe a localized surface-plasmon effect in an AgO_x film that was observed experimentally and calculated numerically with a finite-difference time-domain (FDTD) computer program [4]. The localized surface plasmons lead to an enhanced evanescent field in the vicinity of the scattering center. The nonlinear properties of an AgO_x film were studied by *Ho* et al. with the z-scan technique [6]. Localized surface plasmons were also observed in the aperture-type system by *Tsai* and *Lin* [7]. Radial-direction characteristics of the aperture-type system are reported by *Tominaga* et al. [8]. Pulsed-laser nonlinear properties of both the aperture system and the light-scattering-center system are described by *Fukaya* et al. [9]. Transmission properties of the aperture-type system were studied by *Nakano* et al. [10]. The angular spectrum produced by the light-scattering-type nonlinear layer is reported by *Nakano* et al. through experiment and FDTD simulation [11]. Unusual angular scattering characteristics were observed. An investigation of the super-RENS structure using an FDTD code is also reported by *Peng* [12].

In this paper, we investigate how focused light scatters from a recorded pattern of data marks in a dynamic optical disk with a nonlinear AgOx layer. Observations were made in the pupil of the objective lens. This work expands on the data available in [13] by measuring the angular distribution of signal power in the pupil of the readout optical system. Also, experimental results are compared to both a scalar model and FDTD results. In the following sections, we first review the light scattering properties of optical disks with respect to the angular spectrum of the reflected and transmitted light. The scalar simulation model and results are then discussed in detail. Results from an FDTD simulation are also presented. Next, the experimental setup is described, which is followed by the experimental results and our conclusions.

2 Signal Power in the Angular Spectrum

The focus spot can be decomposed into a collection of plane waves [14]. Interaction with the mark pattern modifies the plane-wave angular distribution, both in transmission and reflection. The pattern of the reflected light at the objective lens is called the *angular spectrum* of the reflected light, because the field amplitude versus position in the pattern corresponds to the amplitude versus angle of plane waves reflected from the surface. A similar angular spectrum is associated with the transmitted light. By understanding the behavior of the reflected and transmitted angular spectra, we can determine useful system characteristics, such as the frequency response, and other aspects of the spot-media interaction.

The classical limit to data mark size on the recording surface is determined by the optical-system frequency response. The spatial frequency is $1/T$, where T is the period of the data mark pattern. As the period decreases, the spatial frequency increases. The frequency response can be understood simply by recognizing the behavior of the reflected light and how the reflected light is collected by the objective lens. For example, Fig. 2 shows the reflected-light distribution for a periodic pattern of data marks along a track. The reflected light consists of three diffracted cones. The direct reflection is the central

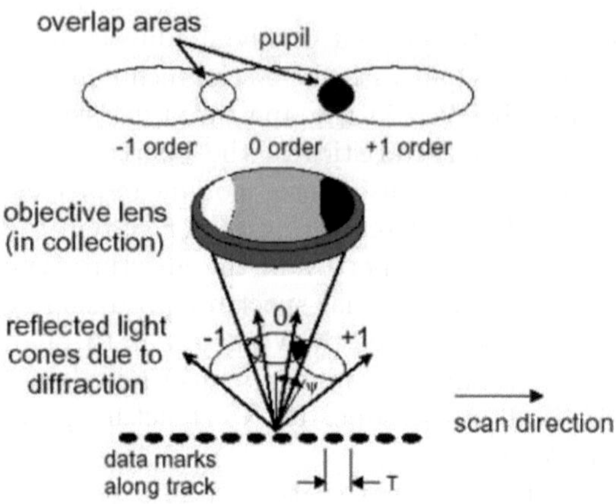

Fig. 2. Overlap of diffracted orders from light reflected off a periodic pattern of data marks along a track causes signal modulation in the angular spectrum at the objective lens

cone. The two outer cones are diffracted orders, which are similar to the central cone in appearance, but they are spread apart by angle Ψ. As T decreases, Ψ increases, and the diffracted orders spread more widely apart; Ψ is also inversely proportional to the laser wavelength. Shorter-wavelength lasers exhibit smaller Ψ.

When the spot scans over the data marks, the optical phase of each diffracted order changes, but the phase of the central cone does not change. The phase difference between the diffracted orders and the central cone produces a modulation in the overlap area due to interference. That is, as the spot scans over the data marks, the overlap areas get brighter and darker as a function of the relative position between the spot and each mark. The brightness of the central cone does not vary. Therefore, the contrast of the signal modulation received at the detectors is determined by the size of the overlap area. More overlap area produces a higher-contrast data signal. As T decreases, so does the overlap area. At some critical mark period, there is no overlap and, consequently, no signal modulation at the detector. This critical mark period is called the *resolution limit* T_{R} of the optical system. A numerical value for T_{R} is found from

$$T_{\mathrm{R}} = \frac{\lambda}{2\mathrm{NA}} = \frac{s}{2}, \tag{1}$$

where NA is the numerical aperture of the objective lens and $s = \lambda/\mathrm{NA}$ is the full-width-at-$1/\mathrm{e}^2$ ($\mathrm{FW}1/\mathrm{e}^2$) spot size. For a DVD system, $\mathrm{NA} = 0.6$ and $\lambda = 650\,\mathrm{nm}$, so $T_{\mathrm{R}} = 541\,\mathrm{nm}$ ($1/T = 1.8\,\mu\mathrm{m}^{-1}$). In practice, the shortest mark period in a DVD system is about $800\,\mathrm{nm}$ ($1/T = 1.25\,\mu\mathrm{m}^{-1}$) due to noise and other system considerations. The signal modulation in the overlap areas can be measured experimentally [15,16]. After writing a single-frequency mark pattern, the signal modulation is measured by raster scanning a point detector in the pupil next to the objective lens.

However, super-RENS systems do not behave in a classical way. The resolution limit for a super-RENS geometry with a DVD-like illuminator is around $T_{\mathrm{R}} = 200\,\mathrm{nm}$ ($1/T = 5\,\mu\mathrm{m}^{-1}$) [10], which is much smaller than the classical resolution limit. *Tsai* et al. suggest that the scattering region is composed of small Ag particles about 15 nm in diameter, which could scatter light preferentially into small angles near the optical axis [4]. Interaction between the focused light beam and the Ag particles has been called a localized plasmon effect. An alternative way to describe the scattering center is as a solid mask of Ag, which completely obscures the data pattern beneath it. Since the opaque region is smaller than the spot, a super-resolution effect is possible that is similar to self-masking magneto-optic media, [17,13] which exhibit improved frequency response compared to simple linear recording layers [18]. In any case, the interaction of the scattering region, the focus spot and the data pattern can be better understood by determining the behavior of the angular spectra generated by a super-RENS disk.

3 Simulation

Signal power distributions in angular spectra are investigated with two simulation techniques. The first technique is a scalar diffraction calculation, and the second technique is a finite-element time-domain (FDTD) calculation. The scalar model provides a convenient and efficient computation engine, but it is limited in accuracy because it does not include vector effects. The FDTD model is a rigorous solution of Maxwell's equations, and therefore includes vector effects. However, the FDTD model is computationally expensive. The following paragraphs describe results from both models, where we find that the scalar model represents the experimental results surprisingly well at low data-mark frequencies.

3.1 Scalar Model Configuration

The scalar simulation geometry is displayed in Fig. 3, where two mark patterns are shown. A focused laser beam with $\mathrm{NA} = 0.6$ and $\lambda = 0.65\,\mu\mathrm{m}$ generates a spot with $s = 1.08\,\mu\mathrm{m}$. Figure 3a shows a track of data marks with 50% duty cycle and $T = 1.04\,\mu\mathrm{m}$. The spatial frequency of this data-mark pattern is $1/T = 0.96\,\mu\mathrm{m}^{-1}$. A solid $0.16\,\mu\mathrm{m}$ diameter Ag particle is positioned $0.16\,\mu\mathrm{m}$ from the spot center. The mark width is $0.4\,\mu\mathrm{m}$. Four regions of reflectivity are identified as r_{A}, r_{B}, r_{C} and r_{D}. Region r_{A} corresponds to the mark area not under the Ag particle. Region r_{B} corresponds to the area between marks not under the Ag particle. Region r_{C} is associated with the Ag particle between the marks. Region r_{D} is associated with the Ag particle overlapping a mark. The complex reflectivity of each area is displayed

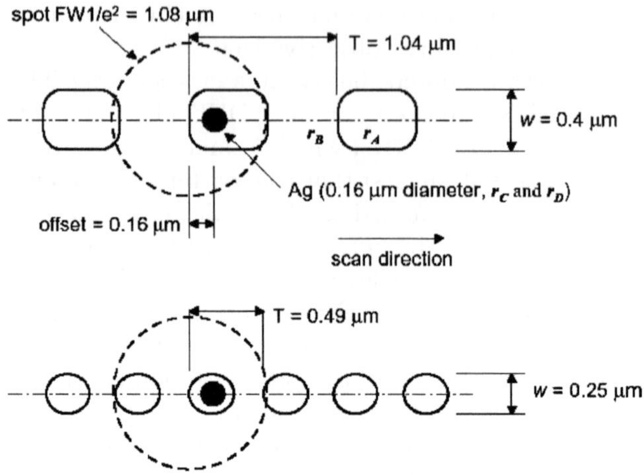

Fig. 3. The scalar simulation geometry for two different recorded mark patterns the super-RENS area is modeled as a constant reflectivity that is delayed by an offset from the center of the focused spot. The reflectivity values $r_{\mathrm{A}} - r_{\mathrm{D}}$ are shown in Table 1. (**a**) Low-frequency pattern; (**b**) high-frequency pattern

Table 1. Reflection and transmission values

Region	Description	$\lvert r\rvert\, e^{j\phi}$	r	$\lvert t\rvert\, e^{j\phi}$	t
A	marks, no Ag	$0.24e^{j0.733}$	0.179 + 0.160j	$0.62e^{-j0.702}$	0.475 – 0.402j
B	between marks, no Ag	$0.29e^{-j0.175}$	0.286 – 0.050j	$0.76e^{-j0.480}$	0.673 – 0.351j
C	between marks, + Ag	$0.56e^{-j1.111}$	0.246 + 0.497j	$0.43e^{-j1.241}$	0.140 – 0.408j
D	marks, + Ag	$0.62e^{-j1.258}$	0.192 – 0.594j	$0.51e^{-j0.922}$	0.309 – 0.408j

in Table 1. Since the magnitudes and phases of regions C and D are similar, the entire area associated with the Ag particle is assigned the reflectivity of region C. The transmission coefficients for each area are also listed in Table 1. Figure 3b displays a mark pattern where $T = 0.49\,\mu\mathrm{m}$ ($1/T = 2.04\,\mu\mathrm{m}^{-1}$). Notice that the mark width is reduced to $T/2$, which is our approximation to the physical mark pattern at this high spatial frequency. The spot size, the diameter of the silver particle and the particle displacement are not functions of T. A third mark period is also investigated, with $T = 0.629\,\mu\mathrm{m}$ ($1/T = 1.59\,\mu\mathrm{m}^{-1}$) and $w = 0.315\,\mu\mathrm{m}$. These three spatial frequencies are chosen to correspond to our experimental values.

4 Scalar Results in Reflection

The average irradiance of the reflected light in the pupil is shown as a contour plot in Fig. 4. The data are scaled with respect to the maximum reflected irradiance in the center of the pattern. The average distribution exhibits a symmetrical Gaussian shape.

Figure 5 shows simulation results for the $1/T = 0.96\,\mu\text{m}^{-1}$ mark pattern in reflection. For Fig. 5a–d, the root-mean-square (rms) signal power is displayed as a function of position in the pupil. Contours of constant rms are displayed relative to the peak rms signal in the pupil. Figure 5a shows the result for scanning with no Ag particle. The signal power is symmetrically distributed in both the scan direction and perpendicular to the scan direction. The bimodal pattern is expected from the overlap of the diffracted orders, as described above. Peak rms signal is listed as −17.5 dB in Table 2, where the dB units are relative to the peak of the average light distribution. That is, a value of −17.5 dB indicates that the peak of the signal distribution in Fig. 5a is a factor of 0.018 lower than the peak of the average irradiance. Also listed in Table 2 is the integrated (sum) rms signal power at −26.4 dB, which is relative to the integrated average light power. Figure 5b shows the result for scanning with a 0.16 μm diameter particle that is offset from the spot center by 0.16 μm. The distribution is changed only slightly, with an asymmetry such that the upper lobe is slightly smaller than the bottom lobe. The presence of the Ag particle reduces the peak signal power by less than 1 dB compared to without the particle for all three cases studied. Figure 5c shows the result for a 0.16 μm diameter particle that is offset from the spot

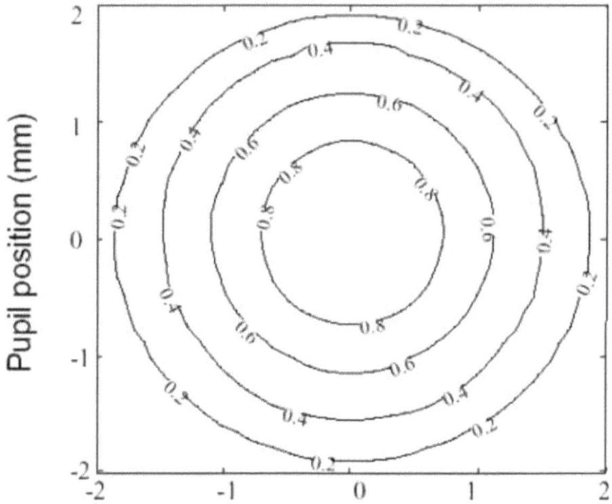

Fig. 4. Average irradiance of reflected light in the pupil as a contour plot. The data are scaled with respect to the maximum reflected irradiance in the center of the pattern. The average distribution exhibits a symmetrical Gaussian shape

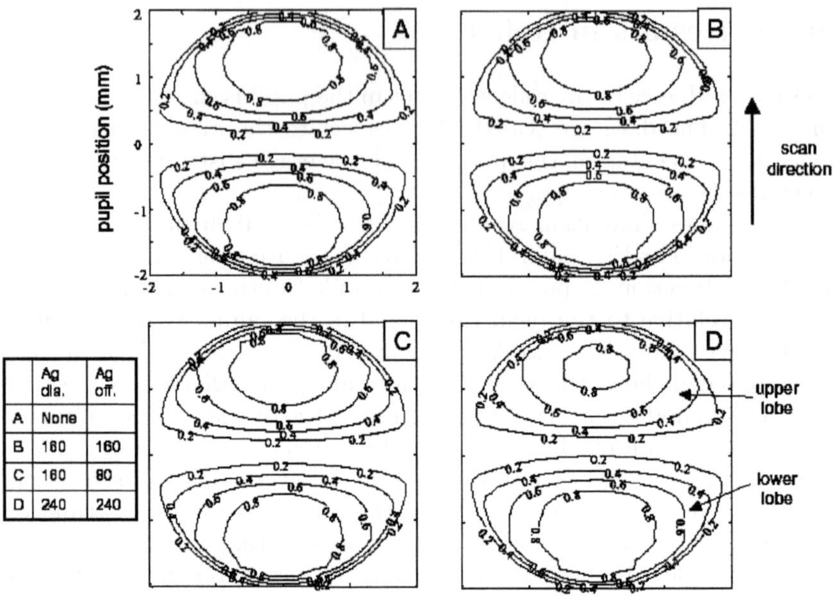

Fig. 5. Scalar simulation results for the $1/T = 0.96\,\mu m^{-1}$ mark pattern in reflection. (**a**)–(**d**) are the root-mean-square (rms) signal powers displayed as a function of position in the pupil. Contours of constant rms are displayed relative to the peak rms signal in the pupil. Peak and integrated signal values are listed in Table 2

Table 2. Reflected light-power ratios for scalar simulation

Spatial frequency (μm^{-1})	Ag diameter (μm)	Ag offset (μm)	dB peak rms	dB sum rms	Figure reference
0.96	240	240	−18.5	−17.5	5D
	160	160	−18.0	−16.7	5B
	160	80	−17.9	−16.7	5C
	none		−17.5	−16.2	5A
1.59	240	240	−31.3	−27.4	6D
	160	160	−32.2	−33.8	6B
	160	80	−31.5	−33.4	6C
	none		−31.4	−40.1	6A
2.04	240	240	−31.2	−31.2	7D
	160	160	−35.8	−35.2	7B
	160	80	−36.3	−34.3	7C
	none		−61.4	−62.0	7A

center by $0.08\,\mu m$. No significant difference is observed when compared to Fig. 5b. In Fig. 5d, the particle diameter is increased to $0.24\ \mu m$ with an offset from the spot center of $0.24\,\mu m$. The asymmetry in the upper lobe is more pronounced.

Figure 6 shows the simulation results for the $1/T = 1.59\,\mu\text{m}^{-1}$ mark pattern in reflection. The peak and integrated signal values are shown in Table 2. Figure 6a shows the result for scanning with no Ag particle. Like Fig. 5a, the pattern is symmetric. However, the overlap of the diffracted orders is extended significantly toward the edge of the pupil, and a much smaller overlap area exists within the pupil boundary. As listed in Table 2, the peak signal level for this case is reduced by an additional 12 dB when compared to the $0.96\,\mu\text{m}^{-1}$ mark pattern. Addition of the Ag particle results in the significantly modified signal shown in Fig. 6b, where a 0.16 µm diameter particle is offset from the spot center by 0.16 µm. Although the peak signal in Fig. 6b is slightly lower than in Fig. 6a, over 6 dB more power exists in the integrated signal. The spread in the signal distribution is most significant for the upper lobe. The lower lobe appears nearly unaffected. A shift of the Ag particle closer to the spot center, as shown in Fig. 6c with a shift of 0.08 µm, spreads the signal power more evenly over the pupil. Again, the shape of the lower lobe is nearly unaffected. Increasing the particle size, as shown in Fig. 6d, reduces the signal spreading and reduces the signal power in the lower lobe. The highest value of the integrated signal is obtained with the largest Ag particle in Fig. 6d, which exhibits a 12.7 dB increase over the case with no Ag.

Figure 7 shows simulation results for the $1/T = 2.04\,\mu\text{m}^{-1}$ mark pattern in reflection. Peak and integrated signal values are shown in Table 2. Figure 7a shows the results for scanning with no Ag particle. Since the signal and integrated signal power ratios are below −60 dB, the signal distribution shown in Fig. 6a is most likely a quantization error. The diffracted orders are well beyond the pupil boundary. However, the addition of the Ag particle redistributes the signal energy so that a significant signal is passed into the pupil. The energy in the signal is distributed primarily in the upper lobe, with a distinct lower lobe appearing only with the largest Ag particle, as shown in Fig. 6d. The peak signal power is approximately 4 dB to 5 dB lower for the smaller Ag particles. Interestingly, the integrated signal power is similar to the $1.59\,\mu\text{m}^{-1}$ mark pattern.

For $1.59\,\mu\text{m}^{-1}$ and $2.04\,\mu\text{m}^{-1}$ spatial frequencies studied with the reflective scalar model, a 240 nm diameter Ag particle exhibits the highest integrated signal power. In addition, the signal power distribution favors the upper lobe. For the lower spatial frequency, the presence of an Ag particle produces only slight variations in the signal distribution.

5 Scalar Results in Transmission

Figures 8–10 display the results of scalar model calculations for the signal power distribution of angular spectra for the super-RENS disk. The signal power ratios are displayed in Table 3. For the most part, the results in transmission are similar to the results in reflection, only the parity of the lobe

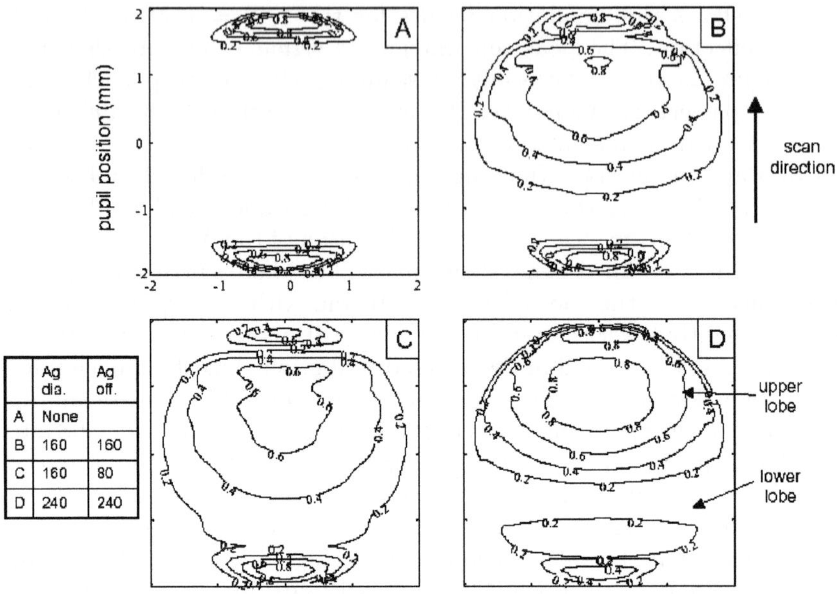

	Ag dia.	Ag off.
A	None	
B	160	160
C	160	80
D	240	240

Fig. 6. Scalar simulation results for the $1/T = 1.59\,\mu\mathrm{m}^{-1}$ mark pattern in reflection. Peak and integrated signal values are shown in Table 2

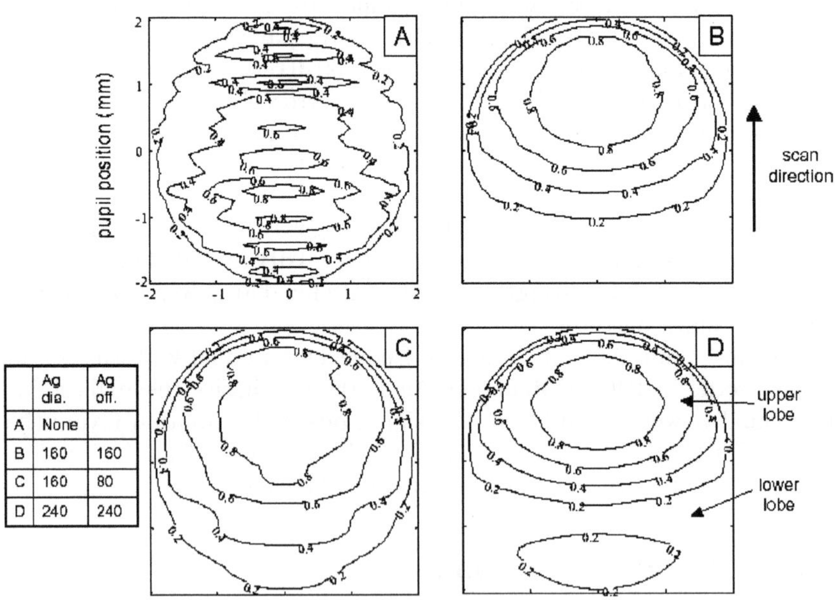

	Ag dia.	Ag off.
A	None	
B	160	160
C	160	80
D	240	240

Fig. 7. Scalar simulation results for the $1/T = 2.04\,\mu\mathrm{m}^{-1}$ mark pattern in reflection. Peak and integrated signal values are shown in Table 2

Table 3. Transmitted light power ratios for scalar simulation

Spatial frequency (μm^{-1})	Ag diameter (μm)	Ag offset (μm)	dB peak rms	dB sum rms	Figure reference
0.96	240	240	−27.6	−26.4	8D
	160	160	−30.5	−29.3	8B
	160	80	−30.8	−29.6	8C
	None		−30.6	−28.9	8A
1.59	240	240	−38.8	−37.7	9D
	160	160	−41.8	−41.6	9B
	160	80	−41.8	−41.0	9C
	None		−41.7	−52.6	9A
2.04	240	240	−42.3	−41.8	10D
	160	160	−46.9	−45.9	10B
	160	80	−46.8	−44.9	10C
	None		−74.3	−72.5	10A

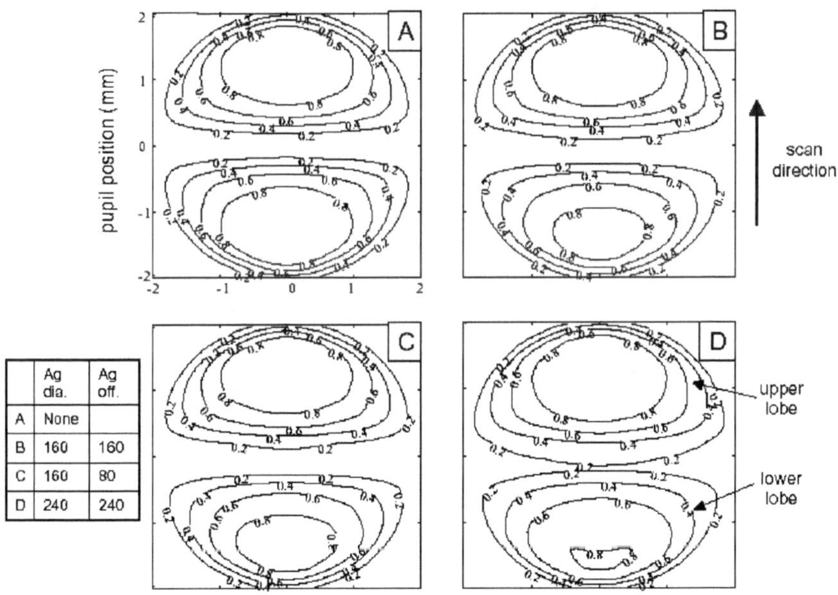

Fig. 8. Scalar simulation results for the $1/T = 0.96\,\mu m^{-1}$ mark pattern in transmission. Figure 5a–d are the root-mean-square (rms) signal powers displayed as a function of position in the pupil. Contours of constant rms are displayed relative to the peak rms signal in the pupil. Peak and integrated signal values are listed in Table 3

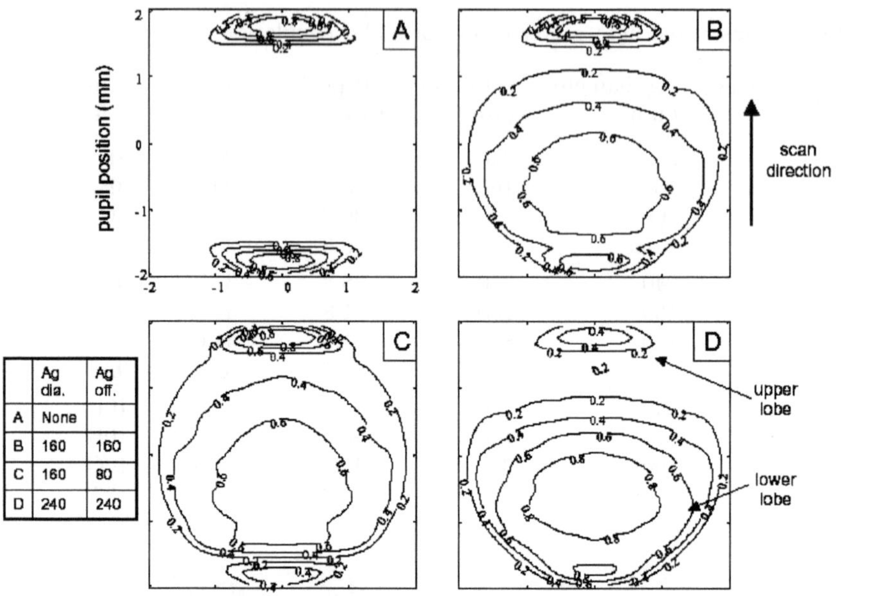

Fig. 9. Scalar simulation results for the $1/T = 1.59\,\mu\text{m}^{-1}$ mark pattern in transmission. Peak and integrated signal values are shown in Table 3

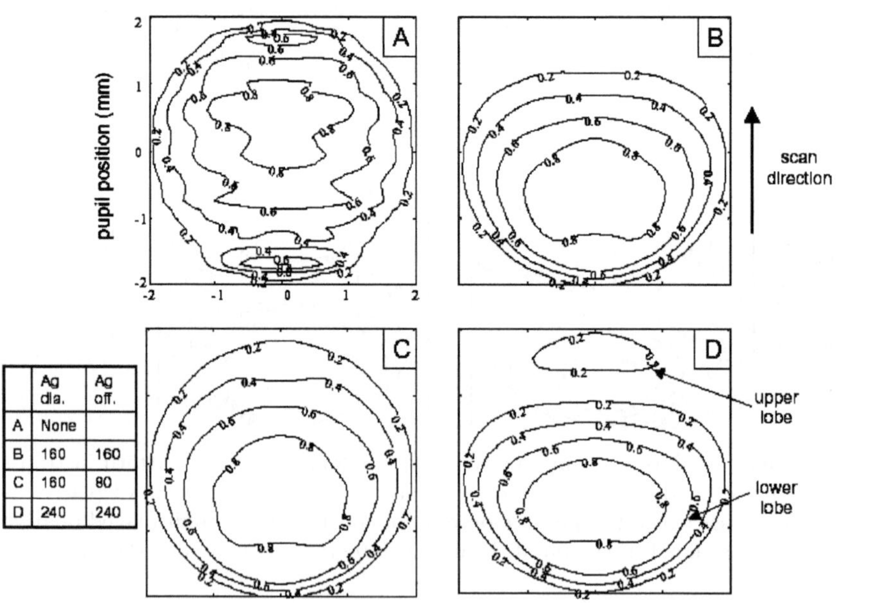

Fig. 10. Scalar simulation results for the $1/T = 2.04\,\mu\text{m}^{-1}$ mark pattern in transimission. Peak and integrated signal values are shown in Table 3

asymmetry is reversed. For example, the lower lobe in the transmitted signal exhibits a stronger dependence on the Ag particle size and position, where the upper lobe displays similar tendencies for the reflected signal. General conclusions about the signal power dependence that are derived from the reflection results are also true for the transmitted signal, except that the signal power in transmission is reduced by 9 dB to 12 dB from the corresponding reflected data.

6 FDTD Results in Reflection

Figure 11 shows a result of an FDTD calculation. In this calculation, $T = 0.36\,\mu\text{m}$ ($1/T = 2.78\,\mu\text{m}^{-1}$) with Ag diameter $= 0.16\,\mu\text{m}$ and offset from spot center $= 0.16\,\mu\text{m}$. Details of the calculation are reported elsewhere [19]. Contours of constant signal amplitude are displayed normalized to the maximum signal amplitude. Pupil radii corresponding to collection numerical apertures of 0.3, 0.6 and 1.0 NA are shown. Three lobes are observed. The lobe in the scan direction is asymmetric in the pupil, and results primarily from the component of incident polarization that is aligned in the scan direction, which is labeled as the x-direction in Fig. 11. This lobe is somewhat consistent with the trend shown in the scalar model at high spatial frequency. The outer two lobes distributed in the direction perpendicular to the scan are nearly symmetric, and they result primarily from the component of incident polarization that is perpendicular to the scan direction. The scalar model does not predict this portion of the scattered angular spectrum.

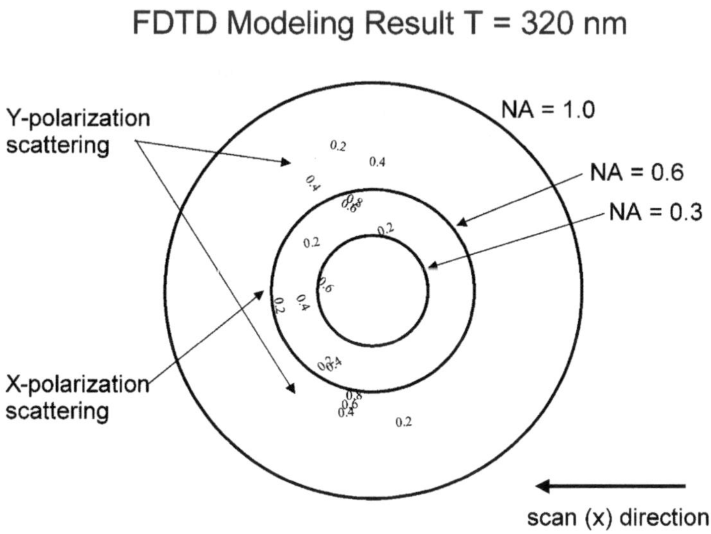

Fig. 11. Result of an FDTD calculation, where $T = 0.36\,\mu\text{m}$ ($1/T = 2.78\,\mu\text{m}^{-1}$) with Ag diameter $= 0.16\,\mu\text{m}$ and an offset from spot center $= 0.16\,\mu\text{m}$

7 Experimental Procedure

The measurement test-stand utilizes a commercial optical recording head that is adapted for ungrooved super-RENS media, as shown in Fig. 12. In the optical recording head, a linearly polarized 0.66 μm wavelength red laser diode is collimated, circularly polarized using a quarter-wave plate, and then focused with a 0.6 NA objective lens through the disk substrate onto the recording layer. The reflected light is collected by the objective lens. The quarter-wave plate repolarizes the light in a linear state that is orthogonal to the original orientation of polarization. The reflected light is then redirected into the servo path (detector A) and the data path (detector B) through a beam-splitter. Detector A is used to control the focus position of the objective lens (30% split) and detector B is used to read the data signal (70% split). Focus correction is performed by moving the objective lens in a voice-coil actuator controlled by the servo-system electronics. The data signal is recovered by sensing a change in the total power of the reflected light at detector B as the recorded data marks pass under the focus spot. Because the disk rotates on a precision air-bearing spindle, concentric data tracks can be written and read without the need for a track-following system.

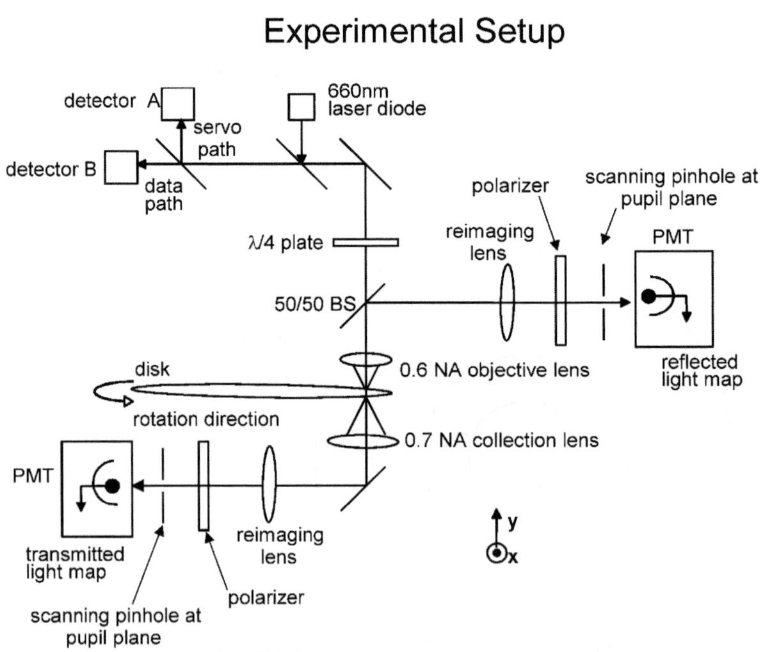

Fig. 12. Experimental setup used to measure signal power in the angular spectrum of the reflected and transmitted data from the super-RENS media. Shown are the components and the beam path through the system

To write a data mark, the focus spot is pulsed to a 9 mW power level as the recording layer moves underneath it at a constant linear velocity of 3.14 m/ s. Tracks of data marks are written with spatial frequencies ranging from $1/T = 0.96$ marks per micron to $1/T = 3.5$ marks per micron. The readout power is between 2.5 mW and 3.5 mW. The data signal detected by detector B is amplified before the signal levels and carrier-to-noise ratio levels are determined.

To map the power distribution of the signal modulation in the reflected angular spectrum, a 50–50 non-polarizing beamsplitter is placed in the beam path between the objective lens and the head, as shown in Fig. 12. The objective lens collimates the reflected light from the disk before it enters the beamsplitter, which redirects the light through the collection optics. The collection optics project an image of the stop at an aperture plane. A scanning 200 µm pinhole is inserted to map out the signal and noise distributions in the x-y plane as a function of position in the aperture plane, which is a 5 mm × 5 mm inscribed circular area. A photomultiplier tube (PMT) behind the pinhole detects the signal. The pinhole and PMT are stepped in 400 µm increments in a rectangular array covering the aperture plane area. At each step, the signal modulation is recorded. Spatial frequencies of $1/T = 0.96$ marks per micron, $1/T = 1.59$ marks per micron, and $1/T = 2.04$ marks per micron are used to generate three independent data sets.

To map the power distribution of the transmitted angular spectrum, a 0.7 NA collection lens is placed directly under the focus spot on the disk, opposite to the optical recording head, as shown in Fig. 12. The scanning pinhole and PMT are inserted close to the aperture of the collection lens to map out the signal and noise distributions in the x-y plane as a function of position. The pupil area is the same as with the reflected aperture plane, i.e. 5 mm × 5 mm. The mapping procedure is the same as is used for the reflected light mapping, using 400 µm steps and the same signal analysis procedure. Also, the same recorded tracks are used to generate the 3D contour maps for the transmitted angular spectrum.

8 Experimental Results

The super-RENS effect is verified by using signals from detector B as the input to a spectrum analyzer and recording the narrow-band carrier-to-noise ratio (CNR) (30 kHz bandwidth) for several single-tone patterns. Figure 13 shows the result of recording spatial frequencies from 1.5 μm^{-1} to 3.8 μm^{-1}. By using detector B, the integrated effect over the pupil is observed. That is, no signal distribution information is included in Fig. 13. However, a strong super-RENS effect is observed. Three curves plotted in Fig. 13 represent CNR for different read power levels. Below 1.5 μm^{-1}, the super-RENS effect is not significant. At 1.59 μm^{-1}, a dramatic 10 dB improvement in CNR is observed when the readout power level at the recording layer is increased

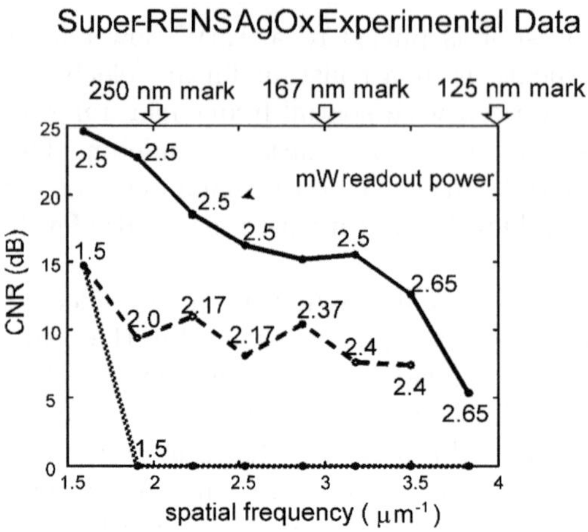

Fig. 13. Spatial frequency versus carrier-to-noise ratio (CNR) of data recorded on super-RENS media with power levels indicated for when the 'aperture opens' under the focus spot in the AgOx layer

from 1.5 mW to 2.5 mW. Below 1.5 mW readout power level, CNR remains at about 15 dB. Higher spatial frequencies exhibit a threshold at which no signal is observable until the power reaches a critical value. CNR drops to zero above 1.59 μm^{-1} for low readout power levels, as indicated by the dotted line in Fig. 13. Threshold CNR is indicated by the dashed line in Fig. 13. The solid line in Fig. 13 represents the maximum CNR obtained in the experiment at each spatial frequency. Notice that mark diameters down to 150 nm can be detected.

The distribution of reflected signal power for a 0.96 μm^{-1} mark pattern is displayed in Fig. 14. Figure 14a displays contours of constant rms signal current relative to the maximum signal current in the scan, which is similar to the simulated result shown in Fig. 5. The signal energy is divided into

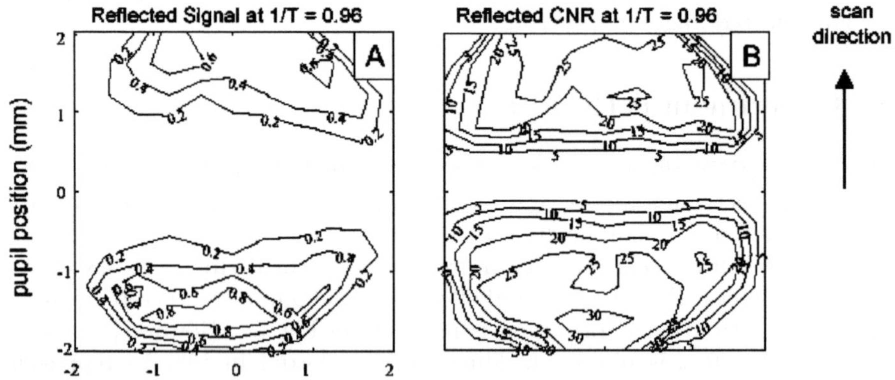

Fig. 14. Distribution of reflected signal power for a 0.96 μm^{-1} mark pattern. (**a**) rms signal amplitude; (**b**) CNR

two lobes along the scan direction. The lobe energy is asymmetric, which indicates that the Ag particle may be larger than 160 nm. CNR data shown in Fig. 14b displays similar characteristics, but CNR data appear more evenly distributed, like the signal distribution in Fig. 5. This appearance may be due to a slightly non-uniform illumination beam, where the signal energy is non-uniform but CNR is not greatly affected. Even with this consideration, the lower lobe exhibits higher CNR than the upper lobe. No significant difference is observed when the polarizer is rotated.

Figure 15 displays the distribution of the reflected signal power for a 1.59 μm^{-1} mark pattern. Figure 15a shows the signal energy when the analyzer is oriented parallel to the scan direction. Asymmetric lobes are observed. The signal distribution indicates that, at this readout power level near threshold, the super-RENS effect does not spread the signal energy as far into the pupil center as predicted by the scalar model in Fig. 6. If the polarizer is rotated across the scan direction, the lobe asymmetry switches and the total energy is reduced, as shown in Fig. 15b. Figure 15c displays the signal without a polarizer and with a higher readout power level near the optimum. Notice that the signal energy is spread significantly farther into the pupil center. In all the cases displayed in Fig. 15, the signal energy of at least

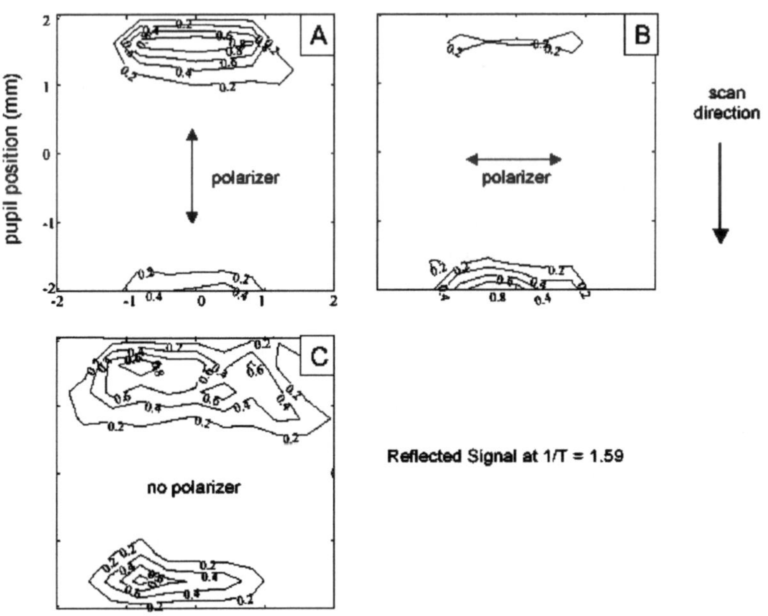

Fig. 15. Distribution of reflected signal power for a 1.59 μm^{-1} mark pattern where the signal energy of at least one lobe extends farther than predicted by the overlap from simple diffracted cones. (**a**) Polarizer aligned with scan direction, and power near threshold; (**b**) Polarizer aligned across scan direction, and power level near threshold; (**c**) no polarizer

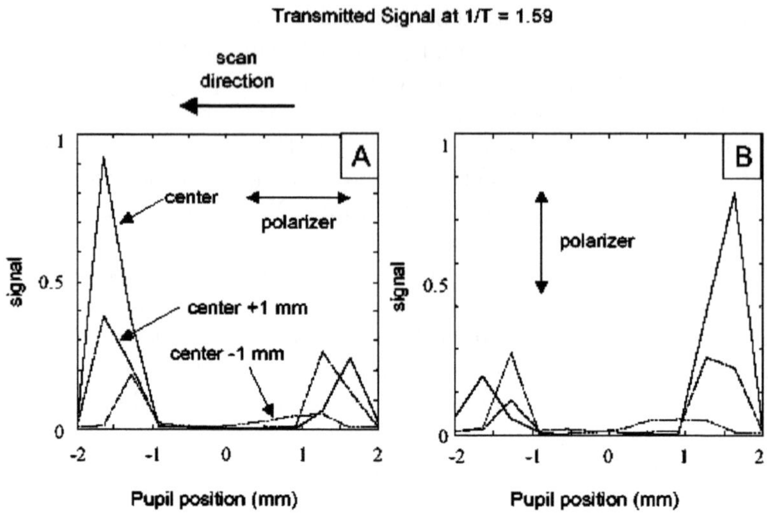

Fig. 16. Profiles of the transmitted signal power across the pupil in the scan direction for $1/T = 1.59\,\mu\text{m}^{-1}$. (**a**) Polarizer aligned with scan direction; (**b**) Polarizer aligned across scan direction. Profiles are measured in the scan direction, with the center profile and two other profiles at $\pm 1\,\mu\text{m}$ from the center

one lobe extends farther than predicted by overlap from simple diffracted cones.

Figure 16 displays profiles of the transmitted signal power across the pupil in the scan direction. Due to difficulty in this transmission measurement, full two-dimensional data maps could not be acquired. Figure 16a shows the signal energy for the polarizer aligned with the scan direction. The lobe energy is asymmetric, as predicted by the scalar model, and the sense of asymmetry is opposite to the reflected data. As with the reflected data, the dominant mode switches when the polarizer is rotated in the cross-scan direction, as shown in Fig. 16b.

Figure 17 shows reflection data for the $2.04\,\mu\text{m}^{-1}$ mark pattern. Figure 16a shows signal-level profiles across the pupil in the scan direction. The profiles are bimodal, and no dominant lobe is observed. The signal power is significantly spread toward the pupil center, but the maximum values remain near the pupil edges in the scan direction. This behavior is not predicted from the scalar results plotted in Fig. 7, where a signal dominant mode appears. In fact, the scalar results suggest bimodal behavior only for a relatively large Ag area (240 nm). The behavior of the signal energy is more like the $0.96\,\mu\text{m}^{-1}$ data, only with lower signal levels. No significant asymmetry is observed, as predicted in Fig. 11 for slightly smaller mark patterns with an FDTD code. Polarization and transmission data are not reported for the $2.04\,\mu\text{m}^{-1}$ mark pattern due to difficulty in obtaining experimental results.

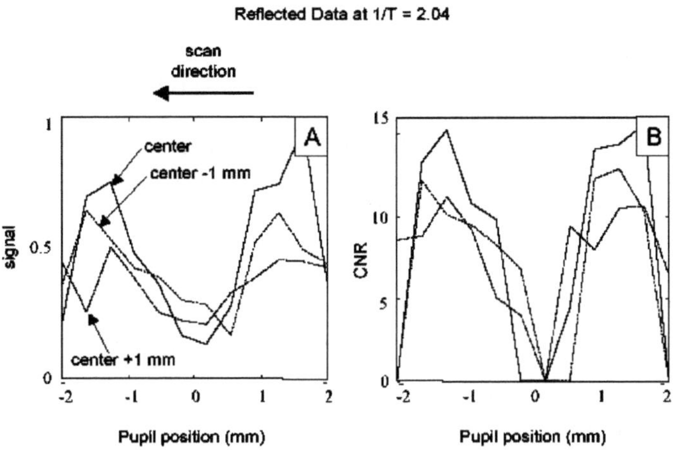

Fig. 17. Reflection data for the 2.04 μm^{-1} mark pattern. (**a**) rms signal amplitude; (**b**) CNR. Profiles are measured as for Fig. 16

9 Conclusions

Both scalar and vector simulations are compared to experimental measurements of signal power in the angular spectrum resulting from mark patterns on super-RENS media. Scalar simulations are fashioned by assuming that the super-RENS effect produces a reflective region that masks data beneath it. Outside this region, the nonlinear layer is transparent. As a focused laser spot scans over the mark pattern, the Ag region moves with the spot, but the center of the region is slightly delayed due to the thermal properties of the multilayer structure. An FDTD vector simulation uses a similar thermal delay and solid Ag area, but full vector interaction of the spot and medium are included. Experimental results are obtained from super-RENS layers on a glass substrate with a dynamic spin stand. Simulated and measured data include signal power distributions at pupil planes both in reflection and in transmission.

Scalar simulation results indicate that an overlapping diffracted-cone argument correctly describes the signal power distribution at 0.96 μm^{-1}, which is below the resolution limit $\lambda/2NA$ for data mark periods. Two signal lobes are distributed in the scan direction. Experimental measurements confirm this result. The only measurable effect of the super-RENS particle is a slight asymmetry in the lobes. In addition, measurements indicate no significant polarization effects in the signal distribution for this data mark period.

For spatial frequencies of 1.59 μm^{-1}, which are near the resolution limit, the super-RENS particle spreads signal energy from the pupil edges toward the pupil center. Two signal lobes are observed in the scan direction, both in scalar simulation and experiment. However, the signal energy is less strongly spread in the experimental observations than what is predicted in the scalar simulation. The optimized readout power level results in the signal energy that is most strongly spread. When a polarizer is rotated from a direction aligned with the scan to orthogonal to the scan, the lobe asymmetry reverses, but no other significant effects are observed.

For spatial frequencies of 2.04 μm^{-1}, which is beyond the resolution limit, the super-RENS particle has a significant effect. The measured results indicate that the signal energy is again distributed in two lobes in the scan direction. These results are not supported by the scalar model, which indicates that the signal energy should primarily be contained in a single lobe. Surprisingly, relatively more energy is directed toward the center of the pupil than what is observed for frequencies near the resolution limit. Perhaps this effect is due to the significant addition of diffracted-cone-overlap energy that is present in the near-resolution-limit results. No diffraction-cone-overlap energy is present in the 2.04 μm^{-1} results. Unfortunately, no polarization-specific data is obtained for this frequency, but the data suggest that the outer lobes present in the FDTD simulation of a higher data frequency are not observed. Also, strong asymmetry in the scan direction predicted by the FDTD simulation is not observed.

In summary, comparison of measured results to a model of the super-RENS effect where the Ag is a thermally delayed Ag area are adequate below the resolution limit and have some similarity near the resolution limit. Above the resolution limit, the super-RENS effect behaves to create signal distributions similar to those observed from lower-frequency mark patterns. No strong lobes are observed in the signal distributions across the scan direction, which indicates that the solid model of the Ag particle may be inadequate to fully explain the behavior for high-frequency mark patterns.

Acknowledgements

The authors would like to acknowledge the support of the Optical Data Storage Center at the University of Arizona and the National Institute of Advanced Industrial Science and Technology (AIST) at Tsukuba-city, Japan.

References

1. J. Tominaga, T. Nakano, N. Atoda: An approach for recording and readout beyond the diffraction limit with an Sb thin film, Appl. Phys. Lett. **73**, 2078–2080 (1998) 119
2. H. Fuji, J. Tominaga, L. Men, T. Nakano, H. Katayama, N. Atoda: A near-field recording and readout technology using a metallic probe in an optical disk, Jpn. J. Appl. Phys. **39**, 980–981 (2000) 119, 120
3. J. Tominaga, H. Fuji, A. Sato, T. Nakano, N. Atoda: The characteristics and the potential of super-resolution near-field structure, Jpn. J. Appl. Phys. **39**, 957–961 (2000) 119

4. W. C. Liu, C. Y. Wen, K. H. Chen, W. C. Lin, D. P. Tsai: Near-field images of the AgO_x-type super-resolution near-field structure, Appl. Phys. Lett. **78**, 685–687 (2001) 120, 123
5. B. J. Bartholomeusz: Thermal response of a laser-irradiated metal slab, J. Appl. Phys. **64**, 3815–3819 (1988) 120
6. F. H. Ho, W. Y. Lin, H. H. Chang, Y. H. Lin, W. C. Liu, D. P. Tsai: Nonlinear optical absorption in the AgO_x-type super-resolution near-field structure, Jpn. J. Appl. Phys. **40**, 4101–4102 (2001) 121
7. D. P. Tsai, W. C. Lin: Probing the fields of the super-resolution near-field optical structure, Appl. Phys. Lett. **77**, 1413–1415 (2000) 121
8. J. Tominaga, H. Fuji, A. Sato, T. Nakano, T. Fukya, N. Atoda: The near-field super-resolution properties of an antimony thin film, Jpn. J. Appl. Phys. **37**, L1323–L1325 (1998) 121
9. T. Fukaya, J. Tominaga, N. Atoda: Nonlinear optical properties of mask layer in super-RENS system, in *Growth, Fabrication, Devices, and Applications of Laser and Nonlinear Materials*, J. W. Pierce, K. I. Schaffers (Eds.), Proc. SPIE **4268**, 79–87 (2001) 121
10. T. Nakano, A. Sato, H. Fuji, J. Tominaga, N. Atoda: Transmitted signal detection of optical disks with a superresolution near-field structure, Appl. Phys. Lett. **75**, 151–153 (1999) 121, 123
11. T. Nakano, T. Gibo, L. Q. Men, H. Fuji, J. Tominaga, N. Atoda: Angular dependence of near-field scattering light from super-resolution near-field structure disk, in *5th Int. Sympo. on Optical Storage (ISOS 2000)*, F. Gan, L. Hou (Eds.), Proc. SPIE **4085**, 201–203 (2001) 121
12. C. Peng: Superresolution near-field readout in phase-change optical disk data storage, Appl. Opt. **40**, 3922–3931 (2001) 121
13. A. Fukumoto, K. Aratani, S. Yoshimura, T. Udagawa, M. Ohta, M. Kaneko: Super resolution in a magneto-optical disk with an active mask, in *Optical Data Storage 1991*, D. B. Carlin, D. H. Kaye (Eds.), Proc. Photo-Opt. Instrum. Eng. **1499**, 216–225 (1991) 121, 123
14. J. W. Goodman: *Introduction to Fourier Optics* (McGraw-Hill, San Francisco 1968) Chap. 3, p. 48 121
15. E. P. Walker: Superresolution applied to optical data storage systems, Dissertation, University of Arizona (1999) 122
16. T. D. Milster, E. P. Walker: Mechanism for improving the signal-to-noise ratio in scanning optical microscopes, Opt. Lett. **21**, 1304–1306 (1996) 122
17. Y. Wu, C. T. Chong: Theoretical analysis of a thermally induced super-resolution optical disk with different readout optics, Appl. Opt. **36**, 6668–6677 (1997) 123
18. T. D. Milster, C. H. Curtis: Analysis of super-resolution in magneto-optic data recorders, Appl. Opt. **31**, 6272–6279 (1992) 123
19. T. Nakano: Polarization dependence analysis of readout signals from disks with small pits beyond the resolution limit, Techn. Dig. Int. ISPS-2001, National Taiwan University (October 2001) 131

Index

Super-Resolution Scanning Near-Field Optical Microscopy

Ulrich C. Fischer[1], Jörg Heimel[1], Hans-Jürgen Maas[1], Harald Fuchs[1], Jean Claude Weeber[2], and Alain Dereux[2]

[1] Physikalisches Institut, University of Münster, 48149 Münster, Germany
fischeu@nwz.uni-muenster.de
[2] Laboratoire de Physique, University of Burgundy, 21004 Dijon, France

1 Introduction

Scanning near-field optical microscopy (SNOM) is a method to obtain information about the optical properties of a sample at a lateral resolution below the diffraction limit of far-field microscopy. In SNOM, a light source of a dimension which is small compared to the wavelength of light and which is held at a small distance from the sample is scanned across the surface of the sample. The modulation by the sample of the light emitted from the source is recorded as a signal. As a general rule one may say that the size of the source and the distance to the sample limit the resolution of SNOM. A radiating self-emitting point dipole may be regarded as an idealized SNOM source. With such a source the resolution of SNOM imaging is expected to be limited by the distance of this dipole to the surface of the object [1]. It is difficult to design a light-emitting SNOM probe corresponding to a dipole at a distance of less than 10 nm from the object and it is therefore difficult to conceive SNOM imaging beyond a resolution of 10 nm. There have been, however, occasional reports of near-field optical imaging at a resolution in the range of 1–10 nm [2,3]. In SNOM-images recorded with the tetrahedral tip (T-tip) a resolution in the range of 1–10 nm was obtained reproducibly on samples consisting of small grains of silver of a size of the order of 2–10 nm embedded in a flat surface of gold [3,4]. An example of an image is shown in Fig. 1. In a different experiment we investigated a surface-embedded latex bead projection pattern [5] consisting of a flat surface of a polymer into which gold patches of a triangular shape of a size of about 50 nm and a thickness of 20 nm were embedded [6]. Characteristic patterns are observed in the SNOM image of the triangular structures at a resolution of 20–50 nm. Using the Green's Dyadic method according to *Girard* and *Dereux* [7] it was possible to interpret these patterns on the basis of a simple model of a light-emitting tip. Herein the tip is a radiating dipole which is located at a distance of 15 nm and an inclination of 45° with respect to the surface of the sample [6]. We defined as photonic nanopatterns such SNOM images as were obtained with

J. Tominaga and D. P. Tsai (Eds.): Optical Nanotechnologies,
Topics Appl. Phys. **88**, 141–153 (2003)

a locally excited dipole as a light-emitting source which can be calculated by numerical methods. The photonic nanopatterns depend on the structure of the object, the orientation and distance of the probing dipole and on the way in which the radiation of the dipole is observed. At first we did not expect that it would be possible to obtain a resolution below 15 nm in such photonic nanopatterns. In the light of the considerations above, the question arises, whether the high resolution obtained in images of the embedded silver grains can be understood by the same model or whether a different mechanism is responsible for image details of a resolution below 10 nm. Here we present SNOM data where a diffuse photonic nanopattern and a highly resolved grain structure of a metal film is observed on the same test sample. A numerical simulation of imaging a model object by a dipole is used in order to judge whether the same mechanism can account for a highly resolved grain structure to appear in a photonic nanopattern in addition to a more diffuse structure.

2 Experimental Scheme

A transmission SNOM/STM (scanning tunneling microscope) setup was used in the experiments of *Heimel* et al. [4], whereas a transmission SNOM/AFM (atomic force microscope) setup was used in the experiments of *Maas* et al. [6]. The STM and AFM modes serve to control the distance between probe and sample and to obtain a topographic image of the sample simultaneously with the SNOM image. A sketch of the SNOM/STM setup is shown in Fig. 2. The T-tip is made of the corner of a glass fragment which is coated with gold. An AFM image of a T-tip is shown in Fig. 3. The grains of the evaporated gold film have a typical diameter in the order of 20 nm. In the SNOM/STM setup (Fig. 2), the fragment is mounted on a small glass prism, such that a beam of light can be focused into the T-tip. Light emitted from the tip is transmitted through the sample and collected with a microscope objective lens. Only the light that passes a pinhole in the image plane is detected by a photomultiplier and used as a SNOM signal. The tip is connected to a sensitive current-to-voltage converter and the tunnel current between tip and sample is used as a signal for the STM mode.

3 Imaging of Photonic Nanopatterns

The test sample consists of a latex bead projection pattern of 20 nm thick gold patches of triangular shape which are embedded in the surface of a polymer film. The polymer film is covered with a 1.5 nm thick film of Pt/C connecting the gold patches by a conductive layer which is necessary for the STM mode [4].

Figure 4a shows the topography of such a sample as recorded in the STM mode. There is no significant topography of the sample on a macroscopic

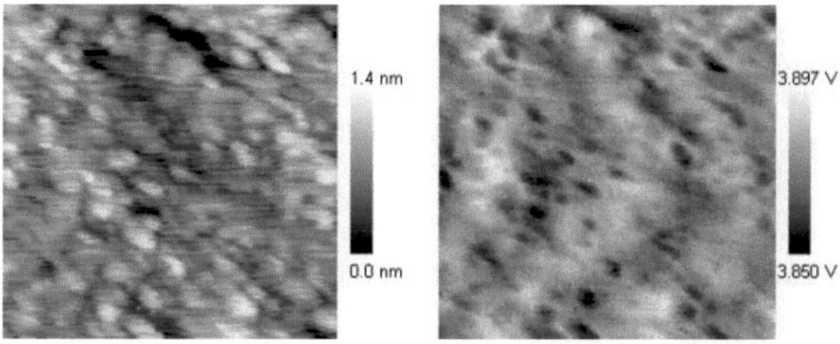

Fig. 1. SNOM imaging of small silver grains (from [4]). The scan range is 125 nm × 125 nm. Topographic image (*left*) obtained by the STM mode and simultaneously recorded SNOM image (*right*) of randomly distributed silver grains embedded into the surface of a thin film of gold. Dark grains in the SNOM image as seen here were interpreted as silver grains embedded into the surface of a flat film of gold [3]

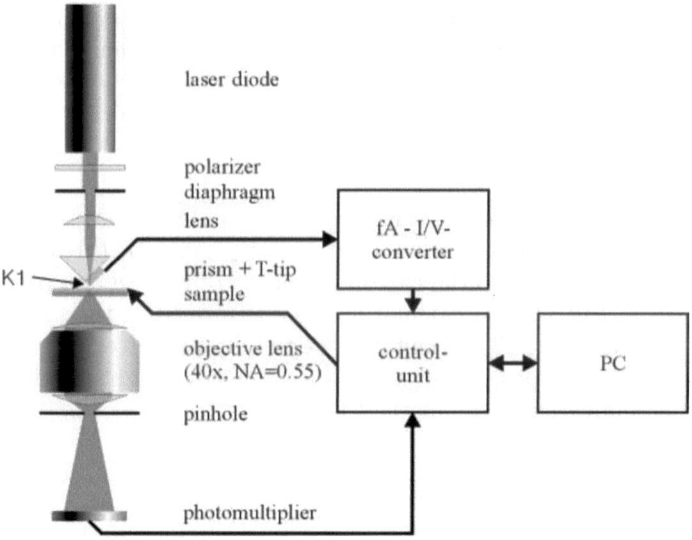

Fig. 2. Scheme of the SNOM/STM setup. The T-tip is mounted on a prism with the edge K1 aligned in the image plane. Light from a laser diode is coupled into the tip. The light emitted from the tip is detected in the image plane of a conventional microscope by a photomultiplier. The T-tip is connected to a sensitive fA-I/V converter

scale, the triangular gold patches are visible only due to the different granular structure of the gold film and the surrounding very thin film of Pt/C. Figure 4b shows the simultaneously recorded SNOM image. A characteristic distribution of bright and dark spots appears in the SNOM image at the location of each triangle, depending on the orientation of the triangles with

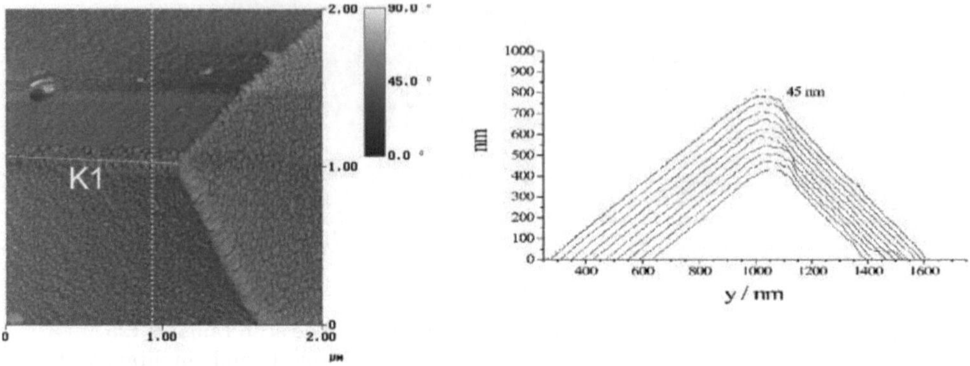

Fig. 3. AFM image of the T-tip. The left image shows the phase signal of an AFM image of the tip, where the granular structure of the gold coating is clearly seen. White lines indicate the edge K1 and the direction of the cross section along which profiles were recorded from the topographic AFM image, as shown on the right-hand side. Assuming that the glass edge K1 is sharp, the extrapolation of the adjacent faces shows that the gold coating on the edge is about 45 nm thinner than on the adjacent faces. This missing thickness corresponds almost completely to the nominal thickness of 50 nm of the gold coating. It is therefore concluded, that the gold coating on the edge K1 is much thinner than on the rest of the glass tip

respect to the T-tip. Similar patterns were observed with a SNOM/AFM setup and were interpreted as photonic nanopatterns [6]. Also, in this case the image shows a close similarity to the calculated image of Fig. 4d of a pair of triangles as shown in Fig. 4. The calculation was performed in the same way as described previously [6]. The orientation of the dipole was chosen to have equal components in the positive y- and z-directions of a right-handed coordinate system, as indicated in the inset of Fig. 4.

4 On the Mechanism of Tip Excitation

As a physical embodiment of the dipole excitation of the T-tip, a dipole induced in a gold grain of a diameter of 20 nm at the apex of the tip may be envisaged. This dipole excitation of the foremost gold grain may act as a local source of light giving rise to the photonic nanopatterns. The mechanism of how the tip is excited is presently not understood in detail, but ideas of how this may occur were described by *Fischer* et al. [8]. A transverse magnetic (TM) edge mode along the glass–metal interface of the edge K1 of the gold-coated glass corner is thought to be excited by the incoming beam of light. Such a TM mode has a longitudinal as well as a transverse electric field component. As shown schematically in Fig. 5, we expect that this mode will be reflected from the corner, leading to a standing-wave excitation of the edge with a node of the transverse component of the electrical field at the tip and a peak of the longitudinal component respectively. In the standing wave,

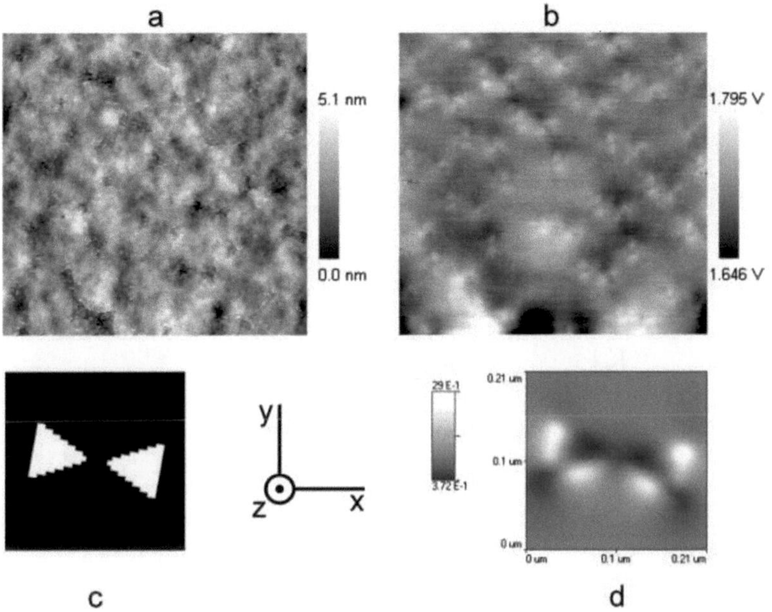

Fig. 4. STM and SNOM image of a test pattern. The test pattern consists of a hexagonal arrangement of gold triangles of thickness 15 nm. (**a**) Topographic image; the scan range is 1 μm × 1 μm. (**b**) Simultaneously recorded SNOM image. (**c**) Model structure used for the calculations; the white triangles are gold triangles of thickness 15 nm. (**d**) Calculated photonic nanopattern of the structure shown in (c), assuming the dipole in the (0,1,1) direction of the (x, y, z) coordinate system indicated in the figure. The orientation of the dipole was chosen to be parallel to the edge K1. All conditions for the calculations were the same as described previously [6]

the transverse components of the waves traveling in opposite directions are expected to cancel each other. Thus the remaining longitudinal component may lead to the dipole excitation of the foremost gold grain oriented parallel to the edge. The orientation of the edges of the T-tip with respect to the image plane of Fig. 4 was known in the experiments. A dipole orientation parallel to the edge K1 of the T-tip as indicated in Fig. 2 was used in the calculations leading to the pattern of Fig. 4d.

5 Highly Resolved SNOM Images of the Granular Structure of a Gold Film

Apart from the diffuse photonic nanopatterns, more details are seen in well-resolved SNOM images of the sample, as shown in Fig. 6 and Fig. 7. Figure 6 shows the topography of a different region of the projection pattern. Due to a defect in the mask of close-packed latex beads used in the fabrication of this sample [5], an extended connected gold patch was created. The SNOM image

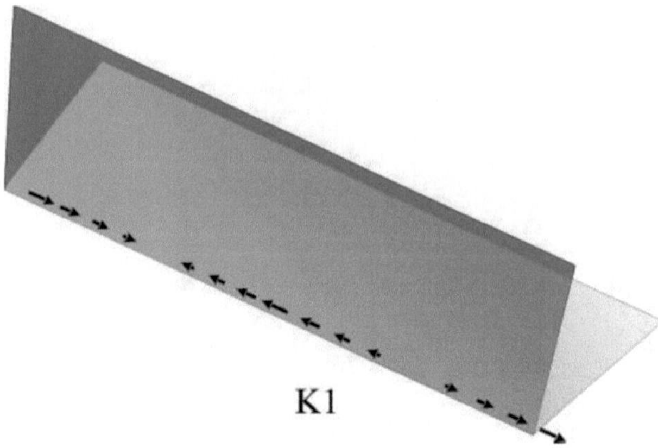

Fig. 5. Schematic view of a longitudinal standing-wave edge mode along the edge K1

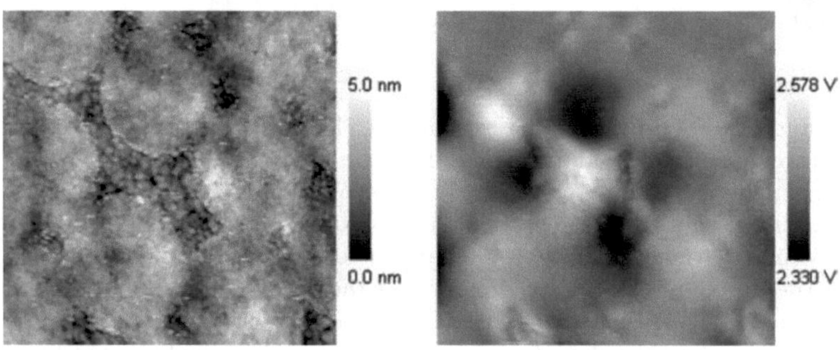

Fig. 6. Topographic STM image (*left*) and SNOM image (*right*) of the test sample. The scan range is 500 nm × 500 nm

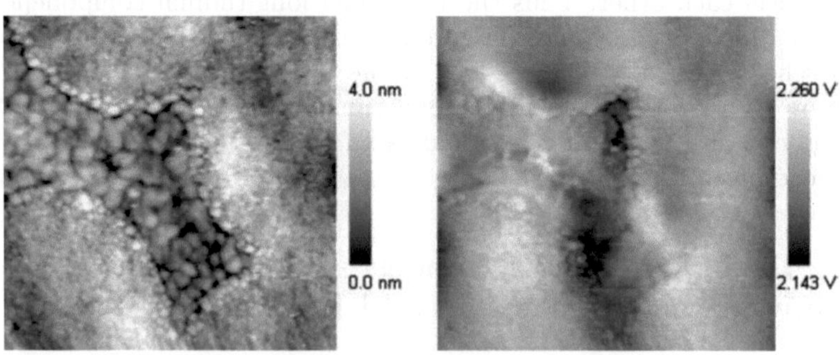

Fig. 7. Zoomed detail of Fig. 6; the scan range is 250 nm × 250 nm. Topographic STM image on the *left*; SNOM image on the *right*

in Fig. 6 shows strong modulations on this structure which have the same origin as the photonic nanopatterns of the triangular patches. Figure 7 shows a zoomed topographic image of the same region, where again the gold grains are clearly resolved. The image reveals – apart from the strong modulation – several other more detailed features. There are some regions where the gold grains are clearly visible, whereas in other regions they remain invisible. Moreover, at different sites the gold grains appear with an inverted contrast. It also appears that at some places closely spaced grains have an inverted contrast at the rim of the structure, where there is a transition from the gold grains to the grains of the Pt/C.

6 Interpretation of the Experimental Images

We conclude that in the SNOM-images of the test structure we observe a photonic nanopattern which can be explained by an obliquely oriented dipole as a source of light for SNOM. On the other hand, the grain structure of the metal film is resolved in some regions of the photonic nanopattern at a resolution below 10 nm. We have the impression that the situation leading to the visibility of the grains in these images is similar to the one where the silver grains were observed. The question arises whether these higher resolved details can be explained by the same model which explains the photonic nanopatterns or whether we have to assume a different mechanism to be responsible for these details. In order to address this question, we performed several calculations with the same model. As is clear from the topographic image of our samples, the gold patches are not homogeneous. They consist of an assembly of gold grains. The test sample was made in such a way that the surface of the gold patches is much smoother on the side which is accessible to the tip than on the side which is embedded in the polymer. The side exposed to the tip has a very small roughness in the order of 4 nm, as is seen from the topographic image. This side is smooth because it is formed as a gold replica of the atomically flat surface of mica, as shown schematically in Fig. 8 [5]. The roughness comes about only because the gold grains do not coalesce completely when the gold film is formed on mica by the evaporation process such that the borders between different gold grains appear as narrow grooves. It is known that gold films tend to grow in a columnar structure [9], and the other surface of the gold film, which is covered by the polymer, should have a very similar granular structure as on the gold mica interface, but the surface of this covered side is much more corrugated. In order to investigate the influence of the granularity on the photonic nanopatterns, we made some model calculations. A 5 nm thick and 100 nm wide quadratic gold slab served as a basic model object, as shown in Fig. 9a. In order to take into account the granularity of the structure we made a second model object of randomly distributed gold grains within the confinement of the same quadratic patch as shown in Fig. 9d. In order to account for the situation that the roughness of our gold samples is larger on the interface between the gold and the polymer than on the surface which is exposed to the tip, we combined the homogeneous patch as a cover layer with the layer of randomly distributed grains, as shown in Fig. 9g. The scanning height of the tip is shown for the different configurations in Fig. 9j. The calculated photonic nanopatterns of the model objects a, d and g are shown below the model objects in Fig. 9b,c, Fig. 9e,f and Fig. 9h,i, respectively, for two different orientations of the dipole within the x-y plane and an equal constant z-component, as indicated by a white line in Fig. 9b,c. Strong modulations are seen in Fig. 9b,c, showing the photonic nanopattern of the homogeneous slab. The appearance of the photonic nanopattern is strongly altered with respect to a homogeneous gold patch for the randomly distributed grains, as shown in Fig. 9e,f. But the resolution of the grains is not better than 20 nm. The composite model object of Fig. 9g accounts for the different roughness of the flat surface exposed to the tip and the other corrugated surface. The photonic nanopatterns of this composite object as shown in Fig. 9h,i show a higher apparent resolution than those of the randomly distributed grains alone of Fig. 9e,f, although in both cases the distance between the tip and the random structure is the same. Although the apparent resolution of the composite film is increased, the grains cannot be clearly recognized in the patterns. In this sense the calculations do not reproduce the experiments where the grains can be clearly identified. But in accordance with the experiments, the numerical calculations also show that higher-resolved details do not appear everywhere in the image and that the orientation of the probing dipole has an important influence where the higher-resolved details appear in the image.

7 Discussion

At first sight it is not clear that a smooth metal cover layer may increase the resolution, because one would assume that the homogeneous cover layer has a shielding effect and thus decreases the contrast as well as the resolution. However, a different consideration may explain in a qualitative way the improved resolution. The scanning dipole is expected to induce a mirror dipole in the cover layer. The electric field distribution of the dipole and its mirror dipole should resemble the field distribution of a quadrupole, which is expected to be much more localized than the field of a single dipole. Such a localization of the near-field may be the origin of higher-resolved details in the SNOM image. The localization is expected to depend on the phase difference between the dipole and its mirror dipole. It is conceivable that the phase relation of the mirror dipole varies with the position of the dipole over the cover layer and that therefore the contrast as well as the resolution varies at different positions of the dipole over the cover layer.

Previous theoretical considerations also give a hint that one may expect an increased resolution in the SNOM images of samples on a metallic substrate

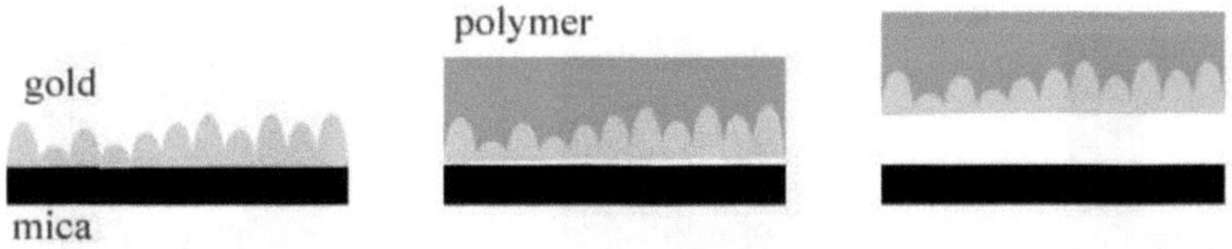

Fig. 8. Schematic view of the structure of the gold film which is formed as a replica of the flat surface of mica and thus has a larger roughness on the face which is covered by the polymer

due to a better confinement of the field when the tip is close to a metallic object: (1) the near-field between a gold sphere (as a model of the tip) suspended above a metal substrate is much more confined than that of an isolated sphere [10]; (2) The electric field between a chain of gold particles supported by a dielectric substrate as generated in a PSTM (Photon Scanning Tunneling Microscope) configuration by a totally reflected beam is much more confined between the spheres than the field around isolated spheres, as was shown experimentally in accord with theoretical calculations by *Krenn* et al. [11]. *Pendry* [12] suggested the use of a thin metal sheet as a "perfect lens" and he claims that by such a device it is possible to overcome the general requirement of close proximity in near-field imaging. For his device rather exotic optical properties of the sheet are required, which cannot be fulfilled in the optical spectral range. But he has also modified these requirements and claims that the property can also be obtained for different conditions in the quasi-static limit for the optical spectral domain. He shows analytically the validity of the concept assuming a grating as a sample for p-polarized evanescent components. The amplification of the evanescent modes by the thin metal sheet seems to be the basic ingredient for this property. It is not clear that the property is also valid for s-polarized evanescent fields and for different structures than linear gratings.

Our experiments show that a resolution enhancement can be expected also for a wider range of optical properties of the metal film. From the experiments it seems that a shielding smooth layer is not essential; adsorption of a very thin object on a flat metal surface or embedding the object into a flat metal surface may be an alternative of a more practical value from the experimental point of view.

There are some aspects of the experimental images which also can be recognized in the numerical images. Higher-resolved details do not appear everywhere in the image. The orientation of the dipole with respect to the object seems to play an important role, where the higher-resolved details appear.

In conclusion, we have demonstrated that the sample as well as the SNOM probe play an important role in the resolution of SNOM imaging. From our results we suggest that a metallic substrate might be important to obtain a resolution in the 1–20 nm range, as obtained experimentally with the tetra-

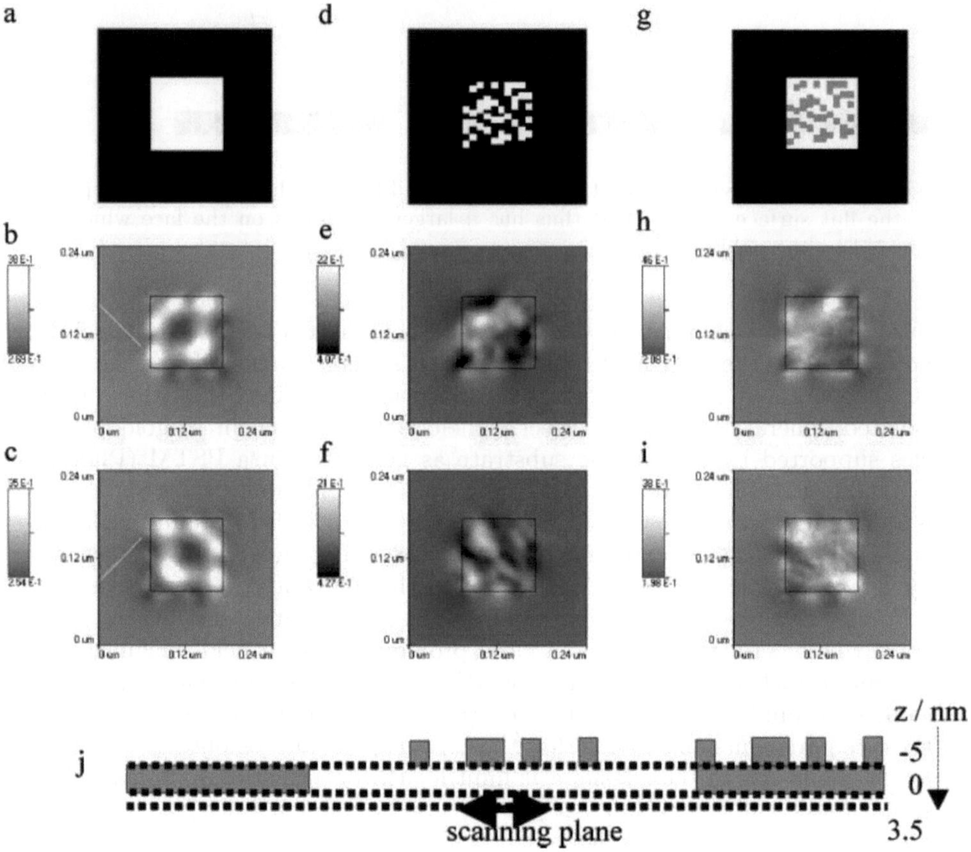

Fig. 9. Model calculations of granular objects. The model objects are shown in (**a**), (**d**), (**g**). A cut showing their position with respect to the scanning dipole is shown in (**j**). The calculated photonic nanopatterns are shown below the model objects in (**b**), (**c**), (**e**), (**f**), (**h**), (**i**). The dipole has a z-component and an equal component in the x-y plane, as indicated by a *white line* in (**b**) and (**c**), respectively

hedral tip. The calculations give a hint that the same mechanism may be responsible for the higher-resolved details in the SNOM images of objects on a metal support and for the more diffuse photonic nanopatterns observed previously. A more detailed comparison between numerical calculations and experimental results is, however, necessary to settle this question. We suggest that SNOM imaging of samples adsorbed on a metal surface is especially attractive for SNOM imaging at high spatial resolution. This is also interesting from the aspect that a high contrast can be expected due to the mechanism scanning tip enhanced contrast [13,14,15,16,17,18].

We think that these conditions will be favorable for the imaging of molecules adsorbed on a metal surface at molecular resolution.

Acknowledgements

We gratefully acknowledge B. Anczykowski from Nanoanalytics for the AFM images of the T-tip of Fig. 3. The research was supported by the German Ministry of Research and Education. UCF acknowledges the Regional Council of Burgundy for supporting a visit at the University of Burgundy.

References

1. J. M. Vigoureux, F. Depasse, G. Girard: Super-Resolution of near-field optical microscopy defined from the properties of confined electromagnetic waves, Appl. Opt. **31**, 3036–3045 (1992) 141
2. F. Zenhausern, Y. Marti, H. K. Wickramasinghe: Scanning interferometric apertureless microscopy, Optical imaging at 10 angstrom resolution. Science **269**, 1083–1085 (1995) 141
3. J. Koglin, U. C. Fischer, H. Fuchs: Material contrast in scanning near-field optical microscopy at 10–1 nm resolution. Phys. Rev. B **55**, 7977 (1997) 141, 143
4. J. Heimel, U. C. Fischer, H. Fuchs: SNOM/STM using a tetrahedral tip and a sensitive current to voltage converter. J. Microsc. **202**, 53–59 (2001) 141, 142, 143
5. U. C. Fischer, J. Heimel, H. J. Maas, M. Hartig, S. Hoeppener, H. Fuchs: Latex bead projection nano-patterns, Surf. Interface Anal. **33**, 75–80 (2002) 141, 145, 147
6. H. J. Maas, A. Naber, H. Fuchs, U. C. Fischer, J. C. Weeber, A. Dereux: Imaging of photonic nanopatterns by scanning near-field optical microscopy. J. Opt. Soc. Am. B **19**, 1295–1300 (2002) 141, 142, 144, 145
7. C. Girard, A. Dereux: Near-field optics theories. Rep. Prog. Phys. **59**, 657–659 (1996) 141
8. U. C. Fischer, A. Dereux, J. C. Weeber: Controlling the light confinement by localized surface plasmon excitation, in *Near-Field Optics and Surface Plasmon Polaritons*, S. Kawata (Ed.), Topics Appl. Phys. **81** (Springer, Berlin, Heidelberg 2001) pp. 49–68 144
9. K. L. Chopra: *Thin Film Phenomena* (McGraw Hill, New York 1969) 147
10. P. K. Aravind, H. Metiu: The effects of the interaction between resonances in the electromagnetic response of a sphere–plane structure: applications to surface enhanced spectroscopy, Surf. Sci. **124**, 506–528 (1983) 149
11. J. R. Krenn, A. Dereux, J. C. Weeber, E. Bourillot, Y. Lacroute, J. P. Goudonnet, G. Schider, W. Gotschy, A. Leitner, F. R. Aussenegg, C. Girard: Squeezing the optical near-field zone by plasmon coupling of metallic nanoparticles, Phys. Rev. Lett. **82**, 2590–2593 (1999) 149
12. J. B. Pendry: Negative refraction index makes a perfect lens. Phys. Rev. Lett. **85**, 3966–3969 (2000) 149
13. J. Wessel: Surface enhanced optical microscopy. J. Opt. Soc. Am. B **2**, 1538–1541 (1985) 150
14. U. C. Fischer: Submicrometer aperture in a thin metal film as a probe of its microenvironment through enhanced light scattering and fluorescence. J. Opt. Soc. Am. B **3**, 1239–1244 (1986) 150

15. U. C. Fischer, D. W. Pohl: Observation of single-particle plasmons by near field microscopy. Phys. Rev. Lett. **62**, 458–461 (1989) 150
16. U. C. Fischer, J. Heimel: Elastic scattering by a metal sphere with an adsorbed molecule as a model for the detection of single molecules by scanning probe enhanced elastic resonant scattering (SPEERS). Jpn. J. Appl. Phys. **40**, 4391–4394 (2001) 150
17. I. S. Averbukh, B. M. Chernobrod, O. A. Sedletsky, Y. Prior: Coherent near-field optical microscopy, Opt. Commun. **174**, 33–41 (2000) 150
18. B. Pettinger, G. Picardi, R. Schuster, G. Ertl: Surface-enhanced raman spectroscopy: towards single molecule spectroscopy. Electrochem. (Japan) **68**, 942–949 (2000) 150

Optical Tunneling Effect and Surface Plasmon Resonance from Nanostructures in a Metallic Thin Film

Wei-Chih Liu

Department of Physics, National Taiwan Normal University,
Taipei, Taiwan 116, Republic of China
wcliu@phy.ntnu.edu.tw

Abstract. Surface plasmon resonance from metallic nanostructures significantly changes the optical response of the material. For metallic thin films with two-dimensional hole arrays or with one-dimensional periodic structures, extraordinary optical transmission has been demonstrated. We report progress on the surface plasmon resonance of periodic narrow-grooved metallic thin films. Two distinguishable modes, surface plasmon polaritons and localized surface plasmons, are associated with different periods. Simulation results demonstrate that the surface plasmon polaritons and localized surface plasmons are coupled with incident electromagnetic waves to resonantly tunnel through the periodic nanostructures. The broadness of localized surface plasmon resonance indicates that it could exist in randomly distributed, narrow-grooved structures as well.

1 Introduction

A flat metallic surface or thin film is usually a very good reflector for visible light. For example, a silver film with thickness t larger than several ten nanometers is effectively opaque. The optical properties of a metal can be demonstrated with a simple jellium model. Consider a metal to be composed of positive ions forming a regular lattice and free conducting electrons. In the jellium model the ionic lattice can be replaced by a uniform positive charge distribution. The plasma frequency ω_p of silver (and most metals) is in the ultraviolet range, and at frequencies below ω_p the dielectric constant ε of silver has a negative real part and only evanescent waves are allowed in the silver.

Subjected to external excitations, the electrons then behave as a gas with a disturbed density and establish longitudinal oscillations of the electronic gas. A metal–dielectric interface may also support charge-density oscillations – surface plasmons – in proper conditions. The surface plasmon excites the electromagnetic field of p-polarized surface oscillations bounded at the interface, and it is essentially a near-field optical phenomenon. The resulting surface electromagnetic wave – the surface plasmon polariton (SPP) – has

J. Tominaga and D. P. Tsai (Eds.): Optical Nanotechnologies,
Topics Appl. Phys. **88**, 153–163 (2003)

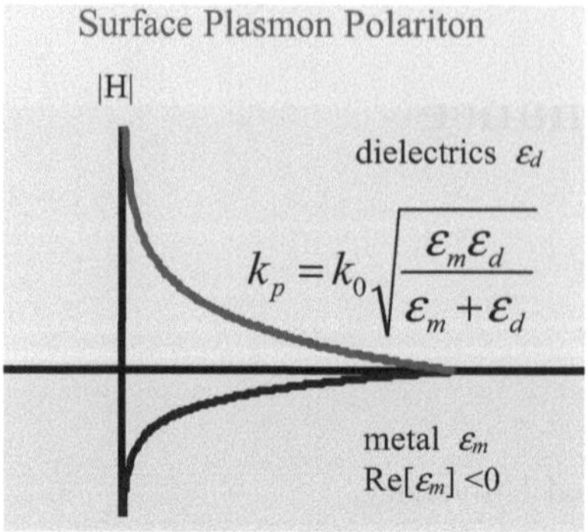

Fig. 1. Surface plasmon polariton at dielectric-metal interface

the form shown in Fig. 1. The parallel (to the interface) component of the propagating constant k_{p} satisfies the equation

$$k_{\mathrm{p}} = k_0\sqrt{\frac{\varepsilon_m\varepsilon_{\mathrm{d}}}{\varepsilon_m+\varepsilon_{\mathrm{d}}}}, \tag{1}$$

where $k_0 = 2\pi/\lambda$ is the propagating constant in vacuum. For example, in the case of the silver–air interface and with incident light of 650 nm wavelength, the relative dielectric constant of the silver is $-19.71 + 0.493j$ and $k_{\mathrm{p}}/k_0 = 1.026 + 0.000685j$. On the other hand, s-polarized surface oscillations do not exist, since the electric field parallel to the interface is continuous across the boundary and thus no surface charge is induced.

A metallic thin film can also support SPP modes. There is no closed-form solution for k_{p} for a thin film, but the dispersion relation can be solved numerically [3]. For a silver thin film in air, Fig. 2 shows k_{p} as a function of the film thickness. When the thickness of the thin film is larger than the skin depth of the silver, k_{p} is close to the solution of the semi-infinite silver–air interface. On the other hand, when the thickness of the thin film becomes smaller, two modes appear – the symmetric mode and the anti-symmetric mode. The value of k_{p} of the symmetric mode approaches the value $k_{\mathrm{d}} = 2\pi n_{\mathrm{d}}/\lambda$ of the surrounding medium, and the value of k_{p} of the anti-symmetric mode approaches infinity as the thickness of the thin film approaches zero [3]. These features are of potential practical interest. The symmetric mode has a small imaginary part and long propagation distances in thin metallic films. For the anti-symmetric mode of thin metallic films, both the real and imaginary parts of k_{p} are large, which is related to evanescent waves with a high propagation constant in a confined area.

Since k_{p} is always larger than the propagating constant in the dielectric in both the interface and the thin-film cases, just shining a light on the dielectric–metal interface cannot excite a SSP. To overcome this phase-matching problem, a periodic metallic structure can provide the in-plane momentum required for the incident light in the appropriate polarization

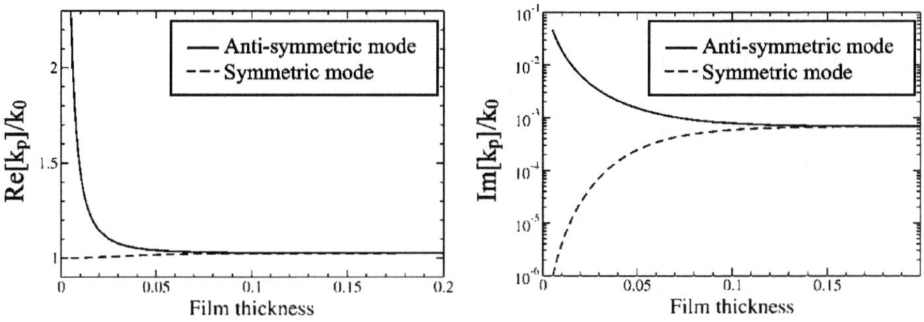

Fig. 2. Real and imaginary part of the normalized propagation constant k_p as a function of the silver film thickness

to excite a SPP, resulting in strong optical absorption. This effect has been extensively studied in the last century [1,2].

An exciting new development in related topics is the study of extraordinarily high optical transmission through metallic films with a periodic array of subwavelength holes by *Ebbesen* et al. [4,5,6,7]. At resonant frequencies, the enhancement of transmission can be several orders of magnitude compared to what is predicted for a subwavelength aperture [8]. The locations of transmission peaks correspond to the SPP resonance modes. Consequently, the phenomenon was explained as plasmon-enhanced light tunneling through subwavelength holes [9,10,11,12].

On the other hand, nanoscale noble-metal structures can exhibit anomalous optical excitation due to the localized surface plasmon (LSP). The LSP resonance positions depend mainly on the structure geometry and the polarization state of the incident light, and the LSP excitation is accompanied with a highly localized field enhancement around the nanostructures. With the fast progress of near-field optics and nano-lithography technology, these properties make such metallic nanoscale structures especially interesting for future micro- or nano-photonic applications.

2 One-Dimensional Theoretical Models

The recent investigation of the surface plasmon resonance in zeroth-order metal gratings, which are gratings with periods smaller than the wavelength of the incident light, is intended to shed light on the physical mechanism of the resonant tunneling of light through thin metal films. Surface plasmon excitation on zero-order gratings by incident TM-polarized radiation (i.e., p-polarized, *H* field parallel to the grating grooves) results in strong field enhancement and absorption. Several researchers studied theoretically a metallic grating with very narrow slits [13,14,15,16] and found that TM-polarized incident light can excite LSP modes in the slits and can lead to large transmission through the grating. However, there is a major difference between a film with an array of subwavelength holes and a film with subwavelength slits, which is that only a subwavelength slit can support propagating modes traveling along the direction perpendicular to the film plane. Therefore, *Tan* et al. proposed to use a silver film with a narrow-grooved zeroth-order grating on both sides to overcome this problem [17]. They found that LSP modes localized in the grooves of the two opposite surfaces, and lead to high transmission for TM-polarized incident light.

We have employed a model similar to the work of *Tan* et al. to show that there are two different mechanisms for enhanced optical transmission through thin metal films with periodic structures at different periods [18]. To reduce the complexity resulting from the metallic dispersion at different incident frequencies, we fixed the wavelength of the incident light at 650 nm and changed the periods instead. Though the resonant modes of a one-dimensional periodic system are not the same as those of a two-dimensional periodic system, it is still important to study the special characteristics of SPP and LSP resonance modes. They appear in much wider areas than the subwavelength hole arrays, and are useful tools to control and selectively transport light.

The model used in our work is a free-standing thin silver film with periodic Gaussian-shaped grooves on both surfaces. As in Fig. 3, the profiles of the left-hand and the right-hand surfaces are symmetric. The depth of these grooves, d, is less than $t/2$, where t is the thickness of the film. The minimum separation between the grooves on the opposite surfaces is larger than zero; therefore, there should be no propagating electromagnetic modes between the surfaces, which is a situation rather more similar to the hole experiment of *Ebbesen* et al. than the narrow slits model. All the silver films have the same groove profile: the depth of the grooves is $d = 50$ nm, and the width of the grooves is $w = 10$ nm. The period, l, between the adjacent grooves, is always less than the wavelength of incident light, $\lambda = 650$ nm. The control variables of the SPP and LSP resonance is the period only, that is, the in-plane momentum provided by the periodic structure.

The two-dimensional finite-difference time-domain (FDTD) method with the perfectly matched layer (PML)[19] is chosen to simulate the optical response, because of its computational advantages of reduced memory require-

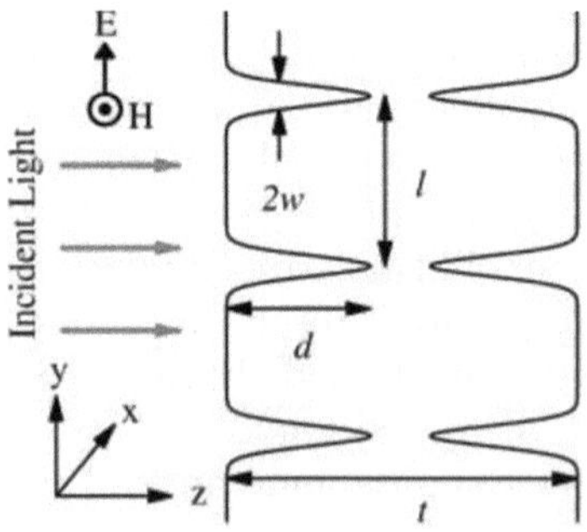

Fig. 3. A schematic illustration of the thin-film grating

ments, freedom from matrix-inversion issues, and ease in treating complex materials and shapes. It allows the calculations of both reflection and transmission of systems, and near-field distributions.

3 SPP and LSP Resonance in the One-Dimensional Deep-Grooved Grating

Figure 4 shows the calculated reflection coefficient and transmission coefficient as a function of the period for silver films with different thicknesses. There are two valleys of the reflection coefficients: the very sharp ones are around $l = 620\,\text{nm}$, and the quite broad ones are at shorter periods. When the thickness of the film decreases, the position of the sharp valley shifts slightly to longer periods, but the broad valley moves significantly to shorter periods. This small shift agrees with the behavior of the symmetric mode of a thin metallic film. Figure 4b shows that a peak in the transmission curve always corresponds to a minimum in the reflection curve at the same period. The sharp transmittance peaks around $l = 616$ to $623\,\text{nm}$ never exceed 0.3, but the broad transmittance peaks can be quite large. This clear contrast suggests that they have different physical mechanisms.

The near-field distributions of the SPP and LSP modes in Fig. 5 show different characters. The narrow transmission peaks shown in Fig. 4b can be identified as the SPP resonance. Figure 5a shows a typical near-field distribution at the SPP resonance. There are strong fields on both sides of the surfaces of the film and the maxima of the fields located at 1/4 and 3/4 period. The near-field distribution agrees with physical pictures of the SPP in a periodic structure. In Fig. 5b there are LSP modes strongly localized in the grooves on both sides of the film surfaces. Resonant excitation of a LSP mode in the left-hand surface grooves results in strong field enhancement. This in turn resonantly excites the LSP mode on the right-hand surface and builds a strong electromagnetic field, which then emits the radiation downwards into the air [17]. This transmission process effectively tunnels light through the metal film via resonantly excited LSP modes.

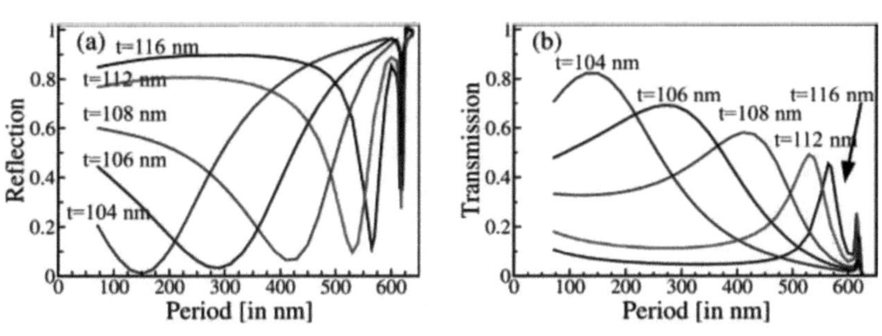

Fig. 4. (**a**)Reflection and (**b**) transmittance coefficients as functions of the period. The thickness of the films is from 104 to 116 nm

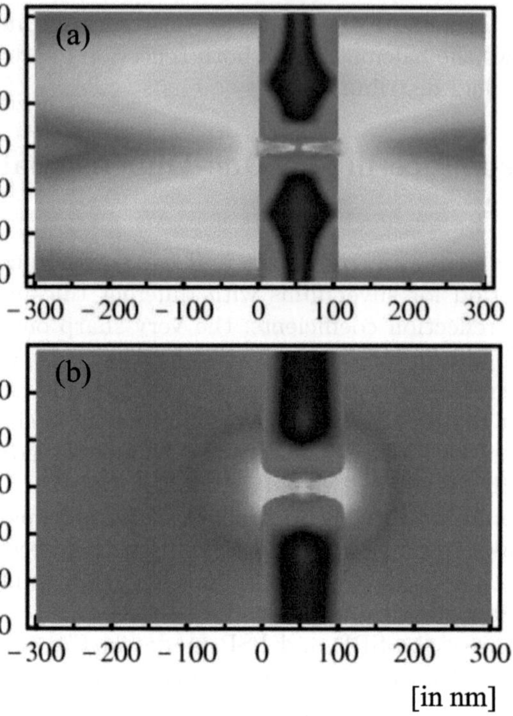

Fig. 5. Near-field distributions of periodic grooves. (**a**) Illustrates the SSP resonance mode for $t = 106\,\mathrm{nm}$ and $l = 621\,\mathrm{nm}$, which is the broad resonant peak of the transmittance curve. (**b**) Illustrates the LSP resonance mode for $t = 106\,\mathrm{nm}$ and $l = 291\,\mathrm{nm}$, which is the narrow resonant peak of the transmittance curve. The field amplitude has been normalized for the incident field amplitude to be 1

An important feature of the general resonance phenomenon is the phase-lag and the time-delay in the resonant response. *Dogariu* et al. [22] have measured the time-delay of a pulse going through the silver film with a subwavelength hole array. At 800 nm resonance, the measured group velocity in the film can be as low as $c/7$. This indicates that the effective index of refraction of the thin film is about 7. Their experiment agrees with a resonantly driven oscillator model and supports the idea that the resonant coupling of light with SPP modes results in the high optical transmission and the slow group velocity of light passing through the silver film with subwavelength hole arrays. To examine the resonant responses in our simulations, we calculated the phase-lag and the effective index of refraction of the outgoing plane wave, comparing with the case without the silver film. The phase-lag for continuous incident light directly corresponds with the time-delay for pulsed incident light. The relation $\Delta\phi = 2\pi(n_{\mathrm{eff}} - 1)\,l/\lambda$ can be applied to obtain the effective index of refraction. In Fig. 6, there is no obvious effective

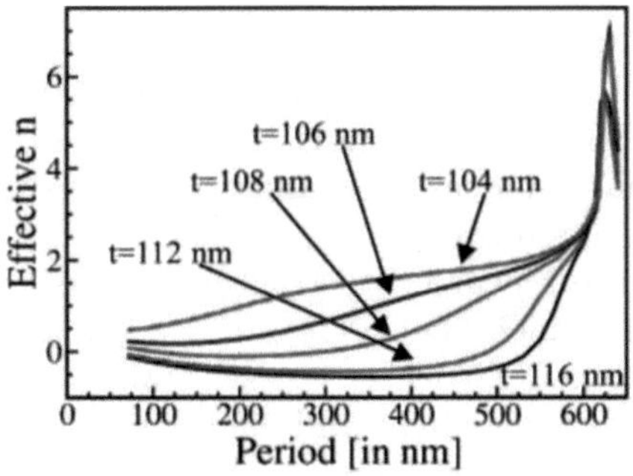

Fig. 6. Effective indexes of refraction as a fuction of period for film thickness $t = 104$–$116\,\mathrm{nm}$

index-of-refraction peak associated with the LSP resonance, except a smooth increase near the LSP resonance position of each thin film with different thickness. On the contrary, the SPP resonance exhibits a large effective index of refraction. At the SPP resonant peak, the effective index of refraction can reach a value as large as 6.9. Though the experiment by *Dogariu* et al. was for a two-dimensional hole array at around 800 nm wavelength and our calculation was for a one-dimensional groove array at 650 nm wavelength, both exhibited a large effective index of refraction for the SPP resonance.

4 Implications for Randomly Distributed Grooves and Non-periodic Grooves

Since the LSP modes are highly localized in the grooves, their optical response curves form very broad bands. The responses of the LSP resonance are not sensitive to the period over a large range of variation. Consequently the broad nature of the LSP resonance can be applied to situations other than periodic arrays of grooves. If the distance between the grooves is slightly different from the resonant period, we can expect the system still to have high transmittance. Therefore, if the groove distances have some small random deviation from the resonant period, we still can anticipate that the system will have high transmittance. This property not only gives large tolerance for future mass-production micro- or nano-photonic devices, but also provides insights into the optical properties of randomly distributed nanostructures.

We calculated the optical response of a thin film to illustrate the optical effect of such randomly distributed grooves. From Fig. 4, the LSP resonant period l is about 290 nm for the thin film of thickness $t = 106\,\mathrm{nm}$, and the peak transmittance coefficient is about 0.69. For this case, we add random deviations in the groove positions from the resonant period. The center positions of the grooves at both sides are now $nl + v \times [2 \times \mathrm{random}() - 1]$, $n = \ldots, -2, -1, 0, 1, 2, \ldots$, where *random*() is a random number from a uniformly random number generator of region (0, 1) and v is the range of the random deviations. For $v = 29\,\mathrm{nm}$ (10% of the resonant period l) the average transmittance coefficient is 0.648 ± 0.027 and for $v = 58\,\mathrm{nm}$ (20% of the resonant period l) the average transmittance coefficient is 0.579 ± 0.019. These are only 94% and 84% of the peak transmittance at the resonant period, respectively, and these examples confirm our hypothesis. The near-field distribution is shown in Fig. 7. Highly enhanced local fields are concentrated in the grooves, similar to the case of periodic grooves.

Another example is the super-resolution near-field structure (super-RENS) [23], which exhibits dynamical nonlinear behavior [24,25,26] and strong local near fields [26,27,28]. The super-RENS has vast potential for applications such as high-density optical storage systems and photonic transistors [29]. The Sb-type super-RENS has very rough interfaces between the 15 nm Sb layer and the protective dielectric layers [28]. The rough interfaces look like randomly distributed narrow grooves and may allow strong LSP resonance when the super-RENS disk is excited with incident light. This hypothesis can explain the strong local near-field and unusual high transmittance of the super-RENS. The very broad nature of the LSP resonance can also explain why the super-RENS respondes effectively to different frequencies of incident light. The physical mechanism of another AgO_x-type super-RENS may also be explained in a similar way. The only difference is that the LSP

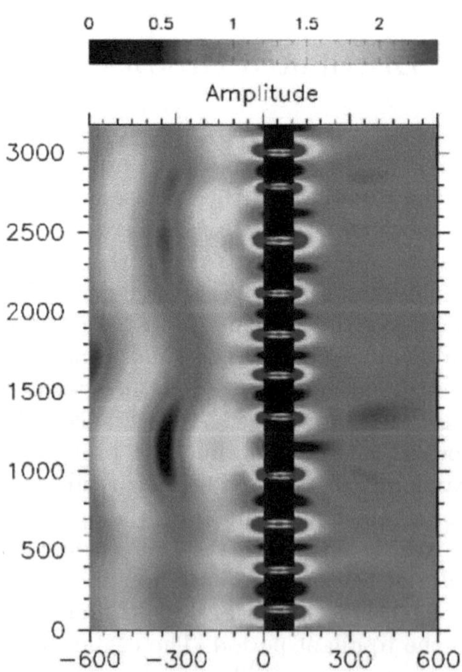

Fig. 7. Near-field distributions of randomly-distributed grooves for $l = 290$ nm and $v = 58$ nm. The thickness of the film is $t = 106$ nm. The field amplitude has been normalized for the incident field amplitude to be 1. The length unit is in nm

resonance is not at small narrow grooves, but at the randomly-distributed nano-sized silver particles embedded in the AgO_x layer [28].

5 Conclusions

These numerical studies on the near-field and far-field optical responses of metallic nanostructures have not only given insight into the physical mechanisms of the unexpected strong light transmission, but they may also lead to micro- and nano-photonics applications. LSP resonance is closely dependent on the details of the nanoscale metallic structures, and similar local field excitation phenomena have often been observed in near-field optical research. The special characteristics of the LSP modes may extend our understanding on the optical properties of periodic and randomly distributed metallic nanostructures. Up to now, little is known about the optical near-fields around nanoscale metal structures, nor about the LSP modes excited by incident light. Better understanding of such phenomena is greatly desirable in order to enable us to manipulate and localize light on the subwavelength scale.

References

1. V.M. Agranovich, D.L. Mills: *Surface Polaritons: Electromagnetic Waves at Surfaces and Interfaces* (North-Holland, Amsterdam 1982) 155
2. H. Raether: *Surface Plasmons* (Springer, New York 1988) 155
3. J.J. Burke, G.I. Stegeman, T. Tamir: Phys. Rev. B **33**, 5186 (1986) 154
4. T.W. Ebbesen, H.J. Lezec, H.F. Ghaemi, T. Thio, P.A. Wolff: Nature **391**, 667 (1998) 155
5. T. Thio, H.F. Ghaemi, H.J. Lezec, P.A. Wolff, T.W. Ebbesen: J. Opt. Soc. Am. B **16**, 1743 (1999) 155
6. T. Thio, H.J. Lezec, T.W. Ebbesen: Physica B **279**, 90 (2000) 155
7. D.E. Grupp, H.J. Lezec, T.W. Ebbesen, K.M. Pellerin, T. Thio: Appl. Phys. Lett. **77**, 1569 (2000) 155
8. H.A. Bethe: Phys. Rev. **66**, 163 (1944) 155
9. I. Avrutsky, Y. Zhao, V. Kochergin: Opt. Lett. **25**, 595 (2000) 155
10. E. Popov, M. Nevière, S. Enoch, R. Reinisch: Phys. Rev. B **62**, 16100 (2000) 155
11. L. Salomon, F. Grillot, A.V. Zayats, F. de Fornel: Phys. Rev. Lett. **86**, 1110 (2001) 155
12. L. Martín-Moreno, F.J. García-Vidal, H.J. Lezec, K.M. Pellerin, T. Thio, J.B. Pendry, T.W. Ebbesen: Phys. Rev. Lett. **86**, 1114 (2001) 155
13. J.A. Porto, F.J. García-Vidal, J.B. Pendry: Phys. Rev. Lett. **83**, 2845 (1999) 155
14. U. Schröter, D. Heitmann: Phys. Rev. B **58**, 15419 (1999) 155
15. S. Astilean, P. Lalanne, M. Palamaru: Opt. Commun. **175**, 265 (2000) 155
16. Y. Takakura: Phys. Rev. Lett. 86, 5601 (2001) 155
17. W.-C. Tan, T.W. Preist, J.R. Sambles: Phys. Rev. B **62**, 11 134 (2000) 156, 157

18. W.-C. Liu, D. P. Tsai: Phys. Rev. B **65**, 155 423 (2002) 156
19. A. Taflove: *Computational Electrodynamics: The Finite-Difference Time-Domain Method* (Artech House, Boston, MA 1995) 156
20. J. J. Burke, G. I. Stegeman, T. Tamir: Phys. Rev. B **33**, 5186 (1986)
21. W.-C. Tan, T. W. Preist, R. J. Sambles, N. P. Wanstall: Phys. Rev. B **59**, 12 661 (1999)
22. A. Dogariu, T. Thio, L. J. Wang, T. W. Ebbesen, H. J. Lezec: Opt. Lett. **26**, 450 (2001) 158
23. J. Tominaga, T. Nakano, N. Atoda: Appl. Phys. Lett. **73**, 2078 (1998) 160
24. T. Fukara, J. Tominaga, T. Nakano, N. Atoda: Appl. Phys. Lett. **75**, 3114 (1999) 160
25. J. Tominaga, H. Fuji, A. Sato, T. Nakano, N. Atoda: Jpn. J. Appl. Phys. **39**, 957 (2000) 160
26. H. Fukaya, D. Buechel, S. Shinbori, J. Tominaga, N. Atoda, D.P. Tsai, W. C. Lin: J. Appl. Phys. **89**, 6139 (2001) 160
27. D.P. Tsai, W. C. Lin: Appl. Phys. Lett. **77**, 1413 (2000) 160
28. W.-C. Liu, C.-Y. Wen, K.-H. Chen, W. C. Lin, D. P. Tsai: Appl. Phys. Lett. **78**, 685 (2001) 160, 161
29. J. Tominaga, C. Mihalcea, D. Buechel, H. Fukuda, T. Nakano, N. Atoda, H. Fuji, T. Kikukawa: Appl. Phys. Lett. **78**, 2417 (2001) 160

Coherent Spontaneous Emission of Light Due to Surface Waves

Jean-Jacques Greffet[1], Remi Carminati[1], Karl Joulain[1], Jean-Philippe Mulet[1], Carsten Henkel[2], Stephane Mainguy[3], and Yong Chen[4]

[1] Laboratoire d'Energétique Moléculaire et Macroscopique, Combustion, Ecole Centrale Paris, Centre National de la Recherche Scientifique, 92295 Châtenay-Malabry Cedex, France
greffet@em2c.ecp.fr
[2] Institut für Physik, Universität Potsdam, 14469 Potsdam, Germany
[3] CEA/CESTA, 33114 Le Barp, France
[4] Laboratoire de Photonique et Nanostructures, Centre National de la Recherche Scientifique, Route de Nozay, 91460 Marcoussis, France

Abstract. Surface-phonon polaritons produce peaks in the local density of electromagnetic states close to the interface. We show that this affects dramatically the spontaneous emission of light by thermal sources in the near field. We discuss several effects such as coherence properties in the near field, radiative heat transfer at short distances and the design of coherent thermal sources in the near field and in the far field.

1 Introduction

When describing different possible sources of light, lasers are often presented as the typical coherent source of light and thermal sources such as a light bulb are often presented as the typical incoherent source. Indeed, for conventional thermal sources, the spectrum is chiefly given by the Planck function, so that the degree of temporal coherence is close to that of blackbody radiation. Concerning spatial coherence, it has been shown that light accross a planar Lambertian source, at a given wavelength λ, is correlated over a distance on the order of $\lambda/2$ [1]. These two results support the statement that conventional thermal sources have a very low degree of temporal and spatial coherence, and therefore are good approximations of incoherent sources.

In deriving the above results, the surface waves are ignored. This is correct insofar as the far field is concerned. However, study of the phenomenon of emission of light into the near field, (i.e. at distances smaller than the peak wavelength from the source) may reveal unexpected behavior. It has been shown recently that the energy density close to an interface increases by several orders of magnitude and acquires a very narrow spectrum [2]. This effect has been shown to be due to the excitation of surface modes. This

J. Tominaga and D. P. Tsai (Eds.): Optical Nanotechnologies,
Topics Appl. Phys. **88**, 163–182 (2003)

suggests that heat transfer in the near field may become very important for very small spacings. This is an important feature for most of the devices that are currently being developed to operate at nanoscales [3,4], such as local probe thermal microscopy [5]. The influence of evanescent waves was studied in pioneering works almost forty years ago [6,7], and more recently by *Loomis* and *Maris* [8]. However, the dramatic enhancement due to the resonant excitation of surface modes was ignored. We will show that the radiative flux may be enhanced by several orders of magnitude due to non-propagating fields, and that it may occur at particular frequencies [9].

A major difference between a laser and a thermal source is that the laser produces a highly directional beam, whereas thermal sources are quasi-lambertian. It has been shown recently [10,11] that the excitation of surface waves modifies dramatically the spatial coherence of surface waves. As a consequence, directional thermal sources or, in other words, partially spatially coherent sources can be designed [12]. We will show that it is possible to fabricate a thermal source that emits light in narrow lobes as does a radio antenna. The physical mechanism will be discussed.

2 Spectral Properties of Emitted Thermal Near Fields

In this section, we consider the emission of a thermal source in the near field. The thermal source is modeled using the geometry depicted in Fig. 1. An interface at $z = 0$ separates a vacuum (medium $z < 0$) from a semi-infinite absorbing material (medium $z > 0$), held at a temperature T. The medium is assumed to be isotropic, local and non-magnetic. It is described macroscopically by its complex frequency-dependent dielectric constant $\varepsilon(\omega) = \varepsilon'(\omega) + \mathrm{i}\varepsilon''(\omega)$.

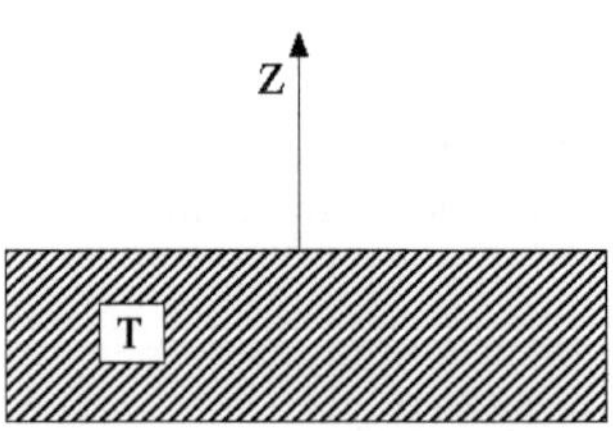

Fig. 1. Geometry of the system used to study thermal emission spectra and near-field spatial coherence. The upper half-space is a vacuum, the lower half-space is a dielectric material with dielectric constant ε at temperature T

2.1 Spectrum of the Thermally Emitted Light

Due to thermal fluctuations inside the material, each volume element behaves as a time-dependent fluctuating dipole. The field radiated by the dipoles through the interface is the thermally emitted field. Note that since we are interested in the properties of the emitted field, we assume that there are no other sources in the problem (there is no incident field from the vacuum side). In particular, this is a non-equilibrium situation. The current density $\boldsymbol{j}(\boldsymbol{r}, t)$

at a point $\boldsymbol{r} = (x, y, z)$ inside the material is a random variable, which is stationary in time [13]. The emitted field $\boldsymbol{E}(\boldsymbol{r}, t)$ is also a random variable. When we take the ensemble average of the field over many realizations of the random currents, we find a zero mean value of the currents and therefore a zero mean value of the emitted field. However, we are interested in the mean value of the density of energy and of the radiative flux. Since these are quadratic quantities, their mean value is not zero. From a more general point of view, the basic quantity to compute is the electric-field cross-spectral density tensor W_{jk} defined by [1]

$$\langle E_j(\boldsymbol{r}_1, \omega) E_k^*(\boldsymbol{r}_2, \omega') \rangle = W_{jk}(\boldsymbol{r}_1, \boldsymbol{r}_2, \omega)\, \delta(\omega - \omega') , \quad (1)$$

where the superscript * denotes the complex conjugate and the brackets denote a statistical ensemble average. The tensor W_{jk} is a measure of the correlation of the field at two different points and at a given frequency. To compute this second-order quantity, we need to know the spatial correlation function of the currents inside the material. It is given by the fluctuation-dissipation theorem. The details of the calculation are given in [10]. The electric energy density $I(\boldsymbol{r}, \omega)$ is proportional to the ensemble average of the square modulus of the electric field. It is deduced from the cross-spectral density by

$$I(\boldsymbol{r}, \omega) = \frac{\varepsilon_0}{2} \sum_{k=x,y,z} W_{kk}(r, r, \omega) . \quad (2)$$

2.2 Examples: Near-Field Thermal Emission of SiC and Glass

Computed thermal emission spectra for silicon carbide (SiC) at different distances z from the interface are shown in Fig. 2 in the frequency range $0 < \omega < 400 \times 10^{12}\ \mathrm{s}^{-1}$ for $T = 300\,\mathrm{K}$. We have used the dielectric constant values given in [15]. In the far field ($z = 100\ \mu\mathrm{m}$), the spectrum is given by the Planck function, except for the frequency $150 \times 10^{12}\ \mathrm{s}^{-1} < \omega < 180 \times 10^{12}\ \mathrm{s}^{-1}$, where SiC is highly reflective, so that it does not emit light. When moving towards *subwavelength* distances from the source plane (the peak emission wavelength at $T = 300\,\mathrm{K}$ is about $10\ \mu\mathrm{m}$), the spectrum changes dramatically. The emission spectrum at $z = 2\ \mu\mathrm{m}$ has a peak at $\omega = 178.7 \times 10^{12}\ \mathrm{s}^{-1}$. At a distance of 100 nm (Fig. 2c), the emission is almost monochromatic, at a frequency not represented in the far-field spectrum. Note that the vertical scale is very different in the three figures. The broad spectrum represented in Fig. 2a has not disappeared, it is merely overwhelmed by the enhancement of the density of energy close to the peak. To summarize, we have two major modifications of the spectral density of energy. First, the spectrum becomes quasi-monochromatic. Second, the electric density of energy is increased by several orders of magnitude. How can we explain these surprising results? The presence of a sharp peak in the emission near-field spectrum can be explained in terms of the density of electromagnetic modes. Indeed, it is known

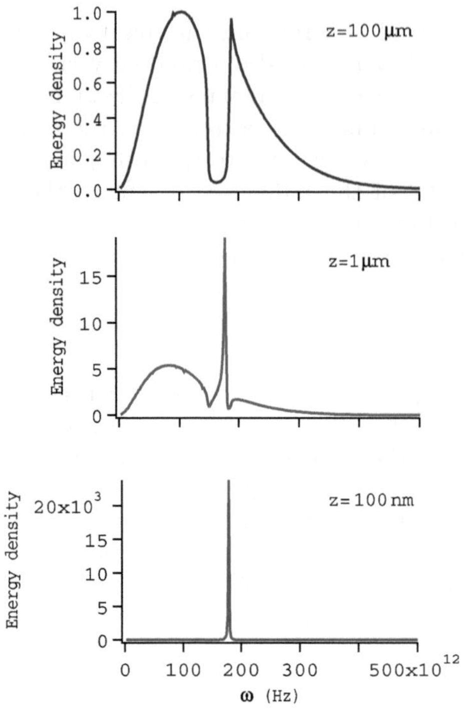

Fig. 2. Thermal emission spectra of a semi-infinite medium of SiC at $T = 300\,\mathrm{K}$, for three different height z above the interface. Note the transition from a broad spectrum to a quasi-monochromatic spectrum as well as the dramatic increase of the electric energy density as z decreases. The data are normalized by the peak value for $z = 100\,\mu\mathrm{m}$

from statistical physics that the density of energy is the product of three terms: the energy of each photon, the density of states and the mean number of photons occupying each state. The latter is given by the Bose–Einstein distribution at equilibrium. Thus, to explain an increase of the density of energy close to the interface, we have to assume that the density of states has been locally increased. The fact that the density of states increases close to the interface is due to the existence of additional solutions of Maxwell's equations: the surface waves. Surface waves are solutions of Maxwell's equations that propagate along the interface and decay exponentially away from the interface. The wave vector $\boldsymbol{k}_{||}$ along the interface is given by [17]

$$\boldsymbol{k}_{||}^2(\omega) = \frac{\omega^2}{c^2}\frac{\varepsilon(\omega)}{\varepsilon(\omega)+1}, \tag{3}$$

where $k_{||} = |\boldsymbol{k}_{||}|$. These solutions exist only when the real part of the dielectric constant is smaller than -1. Thus they exist only for some materials and within certain parts of the spectrum. They correspond to the excitation of coupled mechanical and electromagnetic vibrations. For metals, these are electron density waves and are called surface-plasmon polaritons. For polar dielectrics, they are due to the ion vibrations and are called surface-phonon polaritons. Each surface wave is an additional solution that can host photons. This explains why the energy density increases when ap-

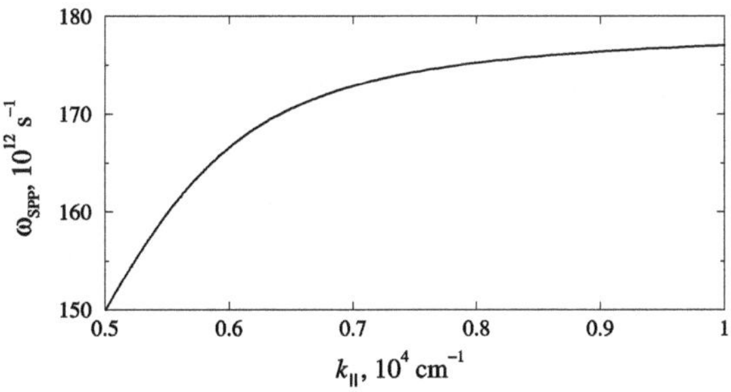

Fig. 3. Dispersion relation of surface-phonon polaritons at a vacuum–SiC planar interface. Re(ω) is plotted versus $k_{||}$

proaching the interface. In order to understand why the spectrum becomes quasi-monochromatic, we have to study in more detail the surface wave. The dispersion relation $\omega(k_{||})$ of the surface wave is shown in Fig. 3. A remarkable feature is that there is an asymptote for a particular frequency. Each couple $(\omega(k_{||}), \boldsymbol{k}_{||})$ is a mode of the problem. Thus, close to the frequency of the asymptote, the local number of modes increases dramatically. This explains the large enhancement of the energy density for a particular frequency and *close to the interface.*

It has been shown in [2] that the electric energy density can be written in the form:

$$I(z,\omega) = \theta(\omega,T)N(z,\omega)\,. \tag{4}$$

In this expression, $2\pi\hbar$ is the Planck constant and $N(z,\omega)$ is the local density of electromagnetic modes (propagating and evanescent) that are excited during the emission process.

The mean energy of the quantum harmonic oscillator in thermal equilibrium at temperature T is $\Theta(\omega,T) = \hbar\omega/[\exp(\hbar\omega/k_BT) - 1]$, where k_B is the Boltzmann constant. It is the product of the energy $\hbar\omega$ of a particle and the mean number of particles per oscillator given by the Bose–Einstein distribution. At short distance z (compared to the peak wavelength of the Planck function), an asymptotic expression of $I(z,\omega)$ can be derived [13]. Upon identification with (4), one obtains

$$N(z,\omega) = \frac{\varepsilon''(\omega)}{|\varepsilon(\omega)+1|^2}\frac{1}{16\pi^2\omega z^3}\,. \tag{5}$$

The $1/z^3$ contribution is a well-known quasi-static property of thermal fields near the source plane [11,13]. It exists for any material and is not related to surface waves. At a given distance z in the near field, the prefactor $\varepsilon''(\omega)/|1+\varepsilon(\omega)|^2$ is responsible for the peak in the spectrum observed in Fig. 2 at $z = 100$ nm. At a frequency $\varepsilon(\omega)_{\max}$ such that $\varepsilon'(\omega_{\max}) \approx -1$, the density of modes displays a sharp peak due to surface modes. Indeed, the peak corresponds clearly to the asymptote of the dispersion relation of the surface waves, as can be seen by comparing with (3). It is also interesting to note that this resonant denominator also appears in the expression of the van der Waals force [14]. Finally, we note that by expanding the real part of the dielectric constant ε' around $\omega_{\max}$ it is easy to show that the peak has a Lorentzian shape with a width given by $[\varepsilon''/(\partial\varepsilon'/\partial\omega)]^{1/2}$.

We show in Fig. 4 a similar behavior for a very common material, namely amorphous glass. In the far field (Fig. 4a), the spectrum is very close to that of a blackbody source. When approaching the surface (Fig. 4b,c), the spectrum changes dramatically and a strong peak emerges. Note that at the location of this peak, the energy density has increased by almost four orders of magnitude. This increase in energy density has important consequences when one considers the radiative heat transfer between two bodies at close distance. This point will be discussed in Sect. 4. The location of this peak also corresponds to a frequency $\omega_{\max}$ such that $\varepsilon'(\omega_{\max}) \approx -1$, and thus to the thermal excitation of resonant surface waves.

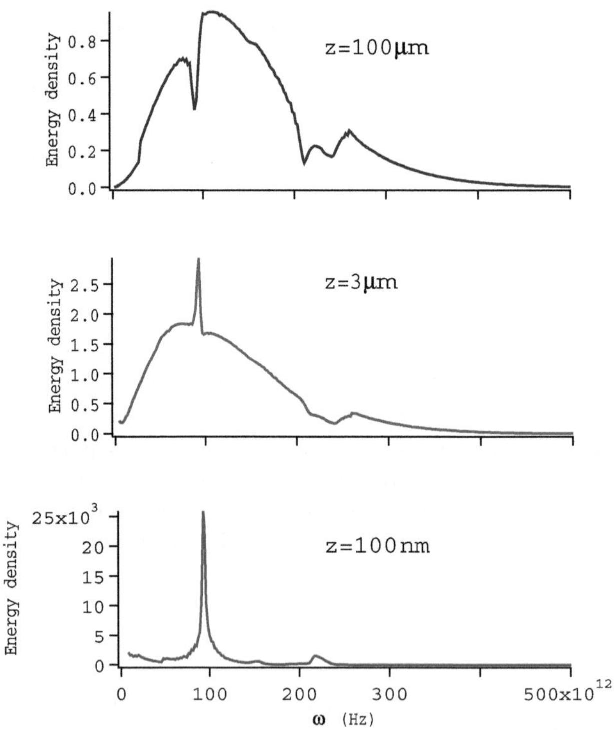

Fig. 4. Thermal emission spectrum of a semi-infinite medium of glass at $T = 300\,\mathrm{K}$, for three different heights z above the interface. The data are normalized by the peak value for $z = 100\,\mu\mathrm{m}$

2.3 Potential Applications

We now turn to the discussion of some implications of the above effects. We shall first discuss the possibility of performing solid-state spectroscopy based on a measurement of the near-field thermal emission spectrum.

Let us assume that a near-field optical measurement yields a signal proportional to the electric energy density[1]. According to (4) and (5), the signal is then proportional to $\varepsilon''/|1+\varepsilon|^2 = 0.5\,\mathrm{Im}[(\varepsilon-1)/(\varepsilon+1)]$. The real and imaginary parts of $(\varepsilon-1)/(\varepsilon+1)$ are linked by the Kramers–Kronig relations (it can be shown that the quantity $(\varepsilon-1)/(\varepsilon+1)$ is causal) so that $\mathrm{Re}[(\varepsilon-1)/(\varepsilon+1)]$ is also obtained. Once these two quantities are known, one can calculate the real and imaginary parts of the complex dielectric constant. In Fig. 5, we have plotted $\varepsilon(\omega)$ obtained by reflectivity measurements on glass [15] (reference curve) and $\varepsilon(\omega)$ deduced from a near-field theoretical spectrum calculated at $z = 100\,\mathrm{nm}$. There is a good agreement between the result and the reference curve. This procedure suggests a new method to perform solid-state spectroscopy on surfaces at a submicrometer scale [2,16]. Note that because

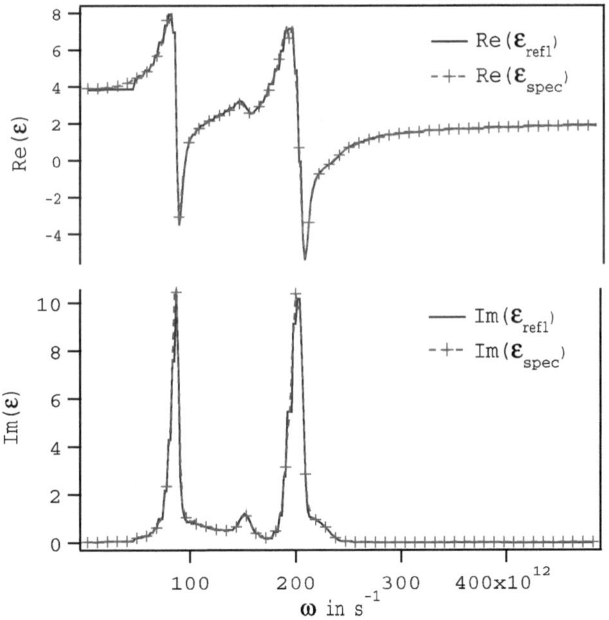

Fig. 5. Real and imaginary parts of the dielectric constant of glass, versus frequency. $\varepsilon_{\mathrm{refl}}$ is the reference value, obtained by reflectivity measurements [15], and $\varepsilon_{\mathrm{spec}}$ is the value numerically reconstructed from the near-field spectrum using Kramers–Kronig relations

[1] In order to account for the polarization dependence of the signal, the correct expression of the signal [18] should be used.

the near-field spectrum is sharply peaked (Fig. 4c), a narrow-band spectrum around the peak frequency is sufficient to get enough information, making it easier to use the Kramers–Kronig relations. A second application could be the use of such thermal sources as near-field infrared quasi-monochromatic sources. Semiconductors display resonance frequencies in the range 2–100 μm. With doped semiconductors, the peak frequency is controlled by the impurity density.

A third application is the increase of the efficiency of thermophotovoltaic generators. These are heaters that burn gas to produce infrared radiation. This radiation is converted into electricity by the photovoltaic effect using low-gap materials. A major drawback of all photovoltaic systems is that the source has a wide frequency spectrum mainly given by the Planck function. Thus, all the photons with energy lower than the energy gap are lost. For photons with energy higher than the gap, the difference $h\nu - E_{\text{gap}}$ is also lost. It is clear that the mismatch between the absorption spectrum and the incident spectrum is the major cause of the reduction of the efficiency of photovoltaic devices. Since the spectrum in the near field becomes almost monochromatic, one may anticipate a tremendous increase in the efficiency of these systems. In what follows, we address the question of the enhancement of the heat transfer due to the excitation of surface-phonon polaritons.

3 Radiative Heat Transfer at Nanometric Distances

We shall now discuss the issue of heat transfer by radiation between closely-spaced bodies.

3.1 Introduction

Besides the application to photovoltaics, there are a number of other fields where heat transfer and temperature control at the submicronic scale is a major issue. For instance, heat transfer in a microprocessor is not very well understood [3]. Thermal probe microscopes with nanometric resolution have been designed [5], and understanding the images is a challenging problem involving nanometer-scale heat transfer. In this section, we study the radiative heat transfer between two semi-infinite bodies separated by a small vacuum gap (Fig. 6). We show that, under certain conditions, the radiative heat transfer due to evanescent waves can be dominant and larger than the classical radiative heat transfer by several orders of magnitude. Moreover, the radiative transfer at short distance may occur at particular frequencies. This is a direct consequence of the near-field spectral behavior discussed in the previous section. Our goal is to investigate how the radiative heat transfer between two plane semi-infinite media depends on the separation distance d (Fig. 6).

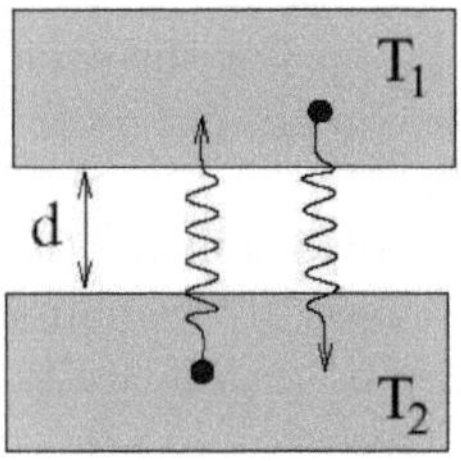

Fig. 6. Geometry of the system used to study nanometer-scale radiative transfer

This problem has already been adressed in the past. *Cravalho* et al. [6] were the first to point out the role of tunneling in radiative heat transfer. *Rytov* [13], *Polder* and *Van Hove* [7] and *Loomis* and *Maris* [8] did a complete calculation based on fluctuational electrodynamics. The radiative transfer is calculated as follows. With reference to Fig. 6, the thermal (fluctuating) currents in medium 1 (temperature T_1) produce an electromagnetic field in all space. The energy dissipated by these fields in medium 2 (temperature T_2) is the first contribution to the radiation heat exchange. An analogous contribution comes from the dissipation in medium 1 of the energy carried by the fields produced by the thermal currents in medium 2. The difference between the two contributions gives the net radiative flux exchanged between both media.

The first contribution to the radiative heat exchange is obtained by calculating the flux of the ensemble average of the Poynting vector $\langle \boldsymbol{S} \rangle = 0.5 \langle \mathrm{Re}(\boldsymbol{E} \times \boldsymbol{H}^*) \rangle$ through a plane just above the interface separating medium 2 and the gap. In this case, $\boldsymbol{E}$ and $\boldsymbol{H}$ are the electric field and the magnetic field in medium 2 generated by the thermal currents in medium 1.

The main points in this calculation are that (1) it accounts for both propagating and evanescent waves (near fields) and (2) the electromagnetic properties of the interfaces are completely taken into account in the electromagnetic model [10]. In particular, the excitation of resonant surface waves is fully described by the poles of the reflection Fresnel factor appearing in the Green's function. In the following, we will show that the contribution of the evanescent waves may enhance the heat transfer by several orders of magnitude when resonant surface waves are involved. We will also see that under these conditions, the radiative transfer becomes quasi-monochromatic.

3.2 Contribution of Resonant Surface Waves

We now evaluate numerically the radiative heat transfer between two half-spaces of absorbing material, separated by a gap with nanometric width d. First of all, let us remark that for most materials, when a gas at atmospheric pressure is present in the gap, the radiative heat transfer remains lower than the conductive heat transfer due to the ballistic flight of molecules between the two bodies [19]. At ambient temperature and pressure and for small distances between the two media, this mechanism yields a conductive heat transfer coefficient[2] h^{C}, which is about $4 \times 10^4\,\mathrm{W\,m^{-2}\,K^{-1}}$ [19]. For the sake of comparison, the value of the heat-transfer coefficient between air and a vertical surface due to natural convection is about $5\,\mathrm{W\,m^{-2}\,K^{-1}}$!

Following the approach of [7], a radiative heat transfer coefficient h^{R} (in $\mathrm{W\,m^{-2}\,K^{-1}}$) can be introduced when $T_1 \approx T_2$. In many cases, h^{R} remains much lower than h^{C}. Nevertheless, we will see that h^{R} may be dramatically enhanced by the thermal excitation of resonant surface waves.

We now consider materials constituting the two media that can support surface waves at a wavelength close to the maximum of the Planck function at 300 K (around $\lambda = 10\,\mu\mathrm{m}$). Many materials such as glass, SiC, and II–VI and III–V semiconductors belong to this category. In Fig. 7 we display h^{R} ($T = 300\,\mathrm{K}$, d) versus the gap width d for SiC and glass. When d is much larger than the wavelength of the maximum of the Planck spectrum (10 µm here), h^{R} does not depend on d. This is the result that would be obtained in classical heat transfer theory, where only propagating waves are taken into account. When d is lower than 1 µm, the heat transfer increases as d^{-2}, in agreement with previous results [8]. Note that for both SiC and glass, h^{R} reaches the typical value of h^{C} at ambient pressure when $d \approx 10\,\mathrm{nm}$. In fact, the radiative heat transfer at small distances is dramatically enhanced by the excitation of resonant surface waves along the interfaces. For example, we have seen that at a distance of 10 nm, h^{R} is four orders of magnitude larger for glass than for chromium, a material which does not support surface waves in the infrared.

The physical origin of the heat-transfer enhancement can be explained from the argument developed in Sect. 2. The electromagnetic energy density above an interface separating a lossy medium from vacuum increases as

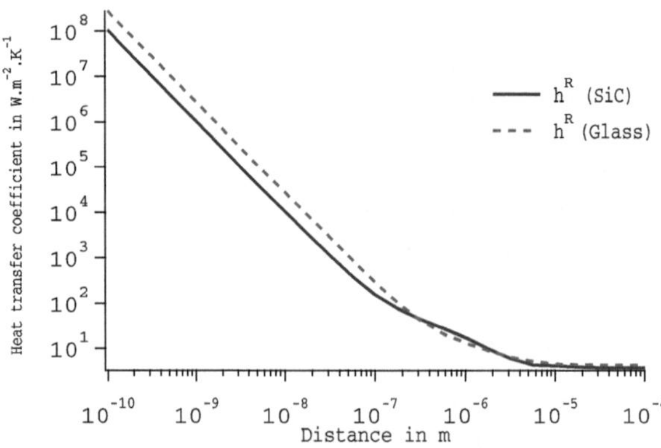

Fig. 7. Radiative heat-transfer coefficient h^{R} at $T = 300\,\mathrm{K}$ versus the gap width d, for two semi-infinite media of SiC or glass

[2] Using this coefficient, the conductive flux per unit surface is written $\phi = h^{\mathrm{C}}(T_1 - T_2)$.

$1/z^3$ at nanometric distances, as shown in (4) and (5). Moreover, at a given distance, the energy density becomes very large at a frequency $\omega_{\max}$ such that $\varepsilon'(\omega_{\max}) \approx -1$. As shown in Sect. 2, this enhancement is related to the existence of additional electromagnetic modes due to the thermal excitation of resonant surface waves. When two media are brought together at a close distance, these evanescent electromagnetic modes contribute to the radiative transfer, through optical tunneling. Near the peak frequency, $\omega_{\max}$, significant radiative energy transfer will take place. A microscopic point of view can be introduced – the heat transfer is due to the coupling between surface photons in the lower and upper interfaces. In other words, when the spacing decreases, there are phonon–phonon collisions.

Because of the existence of this peak frequency, the radiative heat transfer at small distances exhibits peculiar spectral effects. In fact, the frequencies at which the heat transfer occurs strongly depend on the distance d! The increase of the heat transfer coefficients, due to the contribution of evanescent electromagnetic modes, exists at all frequencies but is much larger near the frequency $\omega_{\max}$. To analyze this effect, we can introduce a monochromatic radiative heat transfer coefficient h_ω^{R}. It is plotted in Fig. 8 in the case of SiC and glass, for $d = 10\,\mathrm{nm}$. We see that the heat transfer coefficient exhibits large peaks at particular frequencies. For SiC, the heat transfer is quasi-monochromatic! This is a very unusual situation in heat transfer by radiation between two thermal sources. An asymptotic expansion of h_ω^{R} can be done in the same way as the expansion had been done for the density of modes in Sect. 2. One obtains

$$h_\omega^{\mathrm{R}} \cong \frac{1}{\pi^2 d^2} \frac{\varepsilon_1'' \varepsilon_2''}{|1+\varepsilon_1|^2 |1+\varepsilon_2|^2} k_{\mathrm{B}} \left(\frac{\hbar\omega}{k_{\mathrm{B}}T}\right)^2 \frac{\exp[\hbar\omega/(k_{\mathrm{B}}T)]}{\{\exp[\hbar\omega/(k_{\mathrm{B}}T)]\}^2}\,, \tag{6}$$

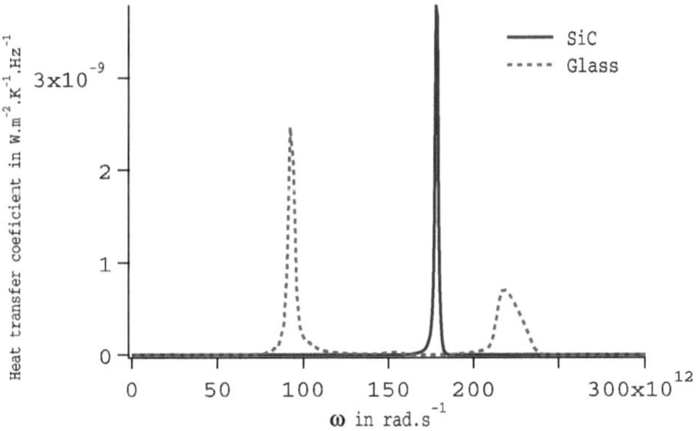

Fig. 8. Monochromatic heat-transfer coefficient h_ω^{R} at $T = 300\,\mathrm{K}$, versus frequency, for two semi-infinite media of SiC and glass; $d = 10\,\mathrm{nm}$. It is seen that most of the transfer takes place in a very narrow range of frequencies

where ε_1'' and ε_2'' are the imaginary parts of the dielectric constants of medium 1 and medium 2, respectively. We see in this equation that the heat transfer is enhanced at the resonant frequencies $\omega_{\max}$ of each material. The enhancement is particularly strong when the materials are identical because the resonances amplify each other. This asymptotic formula also explains why the integrated heat transfer coefficient h^{R} behaves as d^{-2} at small distances in Fig. 7. The monochromatic heat-transfer coefficient h_ω^{R} behaves as d^{-2} in (6). Because it displays narrow quasi-monochromatic peaks, its integral over ω also behaves as d^{-2}. Also note that the temperature dependence of h^{R} is given by (6), using $\omega \approx \omega_{\max}$ in the last term. A behavior very different from the T^4 law of blackbody radiation.

4 Spatial Coherence of Thermal Sources in the Near Field

The study of emission spectra in the near field has shown that thermal emission can be quasi-monochromatic due to the excitation of resonant surface waves. From the point of view of coherence theory, this also demonstrates that such thermal sources exhibit a high degree of *temporal coherence*. From (5) we were able to derive the spectral width of the resonance. The coherence time in the very near field is roughly given by its inverse. It thus strongly depends on the losses at the peak frequency. Again, this large time-coherence in the near field is due to the peak of the local density of states due to the presence of surface waves.

In this section, we will study the issue of the spatial coherence of thermal sources. A well-established result of coherence theory states that light accross a planar Lambertian source (assumed to be a good model for a conventional thermal source), at a given wavelength λ, is spatially correlated over a distance on the order of $\lambda/2$ [1]. In deriving this result, the near-field part of the emitted light is disregarded, because it plays no role in the far-field properties of emission from planar sources. Nevertheless, we have seen in Sect. 2 that the non-propagating (evanescent) fields play a substantial role in the spectral properties of thermal sources. We will see that they also dramatically influence the spatial coherence.

4.1 Exact Calculations of the Spatial Correlation of the Field

When dealing with spatial coherence, we investigate the correlations of the field at different points and equal time. An alternative point of view, is to focus on a particular frequency of the spectrum. It is essential to introduce this frequency analysis because the materials have very different behaviors for different frequencies. Here, we are particularly concerned with the possible presence of surface waves. Therefore, one must study a spatial correlation function of the electric field at a well-defined frequency. In the context of coherence theory, it is called the electric-field cross-spectral density and its definition is given by (1). This quantity can be computed following [10]. We start by showing some typical results obtained close to the interface. Let us first compare the spatial correlation of the field emitted by lossy glass and tungsten, the latter being a metal which does not support surface waves in the visible part of the spectrum. We plot in Fig. 9 the diagonal element $W_{xx}(\boldsymbol{r}_1, \boldsymbol{r}_2, \omega)$ of the cross-spectral density tensor, at a wavelength $\lambda = 2\pi/k = 500\,\text{nm}$. At this wavelength, the dielectric constant [15] of a lossy glass is $\varepsilon = 2.25+0.001\text{j}$ and that of tungsten is $\varepsilon = 4.35 + 18.05\text{j}$. The calculation is performed in a plane $z = z_0$ above the surface of the emitting material. Both $\boldsymbol{r}_1$ and $\boldsymbol{r}_2$ are along the x-axis, and the result is plotted versus $\rho = |\boldsymbol{r}_1 - \boldsymbol{r}_2|$, and normalised by its value at $\rho = 0$. In the very near field ($z_0 = 0.01\lambda$), the curve corresponding to glass (solid line) drops to negligible values after $\rho = \lambda/2$, showing that the correlation length of the x-component of the field is $\lambda/2$. In fact, the solid curve in Fig. 9 strongly resembles the $\sin(kr)/kr$ shape of the cross-spectral density in the source plane of a Lambertian source, previously obtained in the scalar approximation [1].

In comparison, the case of tungsten (dotted curve) is completely different. The correlation length is much smaller than $\lambda/2$ (i.e. much smaller than for the blackbody radiation), on the order of 0.06λ. This subwavelength correlation length is a pure near-field effect, due to non-radiative evanescent fields. At a distance $z_0 = 0.1\lambda$, we see that the correlation length for tungsten (dashed curve in Fig. 9) is much larger (on the order of 0.4λ) than that obtained with $z_0 = 0.01\lambda$ (dotted curve).

We now turn to the study of spatial coherence in light emission from materials supporting resonant surface waves, such as surface-plasmon or surface-phonon polaritons [17]. The thermal excitation of a surface polariton induces some spatial correlation in the field close to the surface, and we may expect a large increase of the correlation length. We illustrate in Fig. 10 the effect of surface-plasmon (Fig. 10a) and surface-phonon (Fig. 10b) polaritons on the spatial coherence of the thermal near field. We plot in Fig. 10a the element

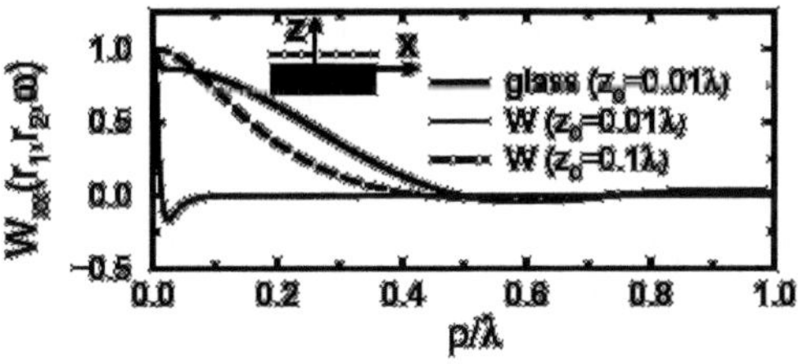

Fig. 9. $W_{xx}(\boldsymbol{r}_1, \boldsymbol{r}_2, \omega)$ in the plane $z = z_0$ versus $\rho = |\boldsymbol{r}_1 - \boldsymbol{r}_2|$; $\boldsymbol{r}_1$ and $\boldsymbol{r}_2$ are on the x-axis; $\lambda = 500\,\text{nm}$. Two materials are considered: lossy glass ($z_0 = 0.01\lambda$) and tungsten ($z_0 = 0.01\lambda$ and 0.1λ). All curves are normalized by their maximum value at $\rho = 0$. Reprinted from [10]

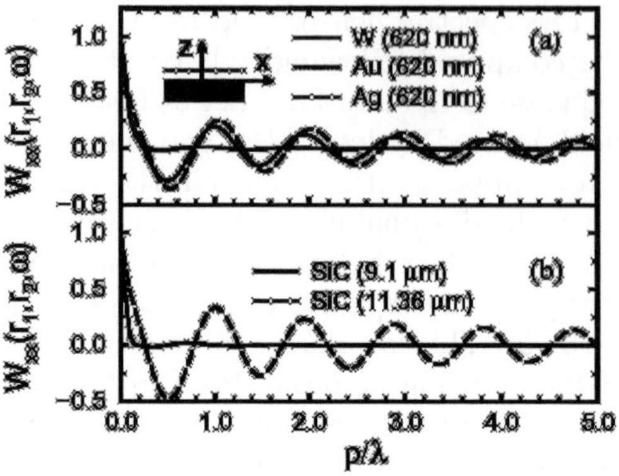

Fig. 10. Same as Fig. 9, with $z_0 = 0.05\lambda$. (**a**) Case of three metals (tungsten, gold and silver), $\lambda = 620\,\text{nm}$. (**b**) Case of SiC with $\lambda = 620\,\mu\text{m}$ and $\lambda = 11.36\,\mu\text{m}$. At 620 nm, the metal dielectric constants are $\varepsilon = 4.6 + 20.5\text{j}$ for tungsten, $\varepsilon = -8.26 + 1.12\text{j}$ for gold and $\varepsilon = -15.04 + 1.02\text{j}$ for silver. Reproduced from [10]

W_{xx} of the cross-spectral density tensor at the wavelength $\lambda = 620\,\text{nm}$, and in the plane $z_0 = 0.05\lambda$, for three different metals.

Both gold and silver exhibit surface-plasmon resonances at this wavelength. We clearly see that whereas the spatial correlation length for tungsten is a fraction of the wavelength (as in Fig. 9), the correlation length for gold and silver is much larger. In fact, although Fig. 10 is limited to $\rho < 5\lambda$ for the sake of visibility, the correlation extends over a larger distance given by the attenuation length of the surface-plasmon polariton. For gold and silver, the attenuation lengths are 16λ and 65λ, respectively. The same effect is seen in Fig. 10b for a SiC crystal, which exhibits a surface-phonon polariton resonance at the wavelength $\lambda = 11.36\,\mu\text{m}$ ($\varepsilon = -7.56 + 0.41\text{j}$) and no resonance at $\lambda = 9.1\,\mu\text{m}$ ($\varepsilon = 1.80 + 4.07\text{j}$). The difference of behavior of this material at the two different wavelengths is striking in Fig. 10b. The correlation length is much higher in the presence of the resonant surface-wave (dashed line) than in the case where no surface wave is excited (solid line). The propagation distance of the surface-phonon polariton in this case is 36λ ($\lambda = 11.36\,\mu\text{m}$).

4.2 Qualitative Discussion

An asymptotic evaluation of the cross-spectral density tensor can be performed in the near field [11]. This analysis allows us to retrieve the previous results analytically and yields physical insight into the mechanisms responsible for near-field spatial coherence effects. In particular, several contributions to the cross-spectral density W_{jk} can be identified: the quasi-static field (extreme near field), surface-phonon or surface-plasmon polaritons, skin-layer currents and far-field contributions. In fact, the asymptotic analysis shows that W_{jk} is the sum of several terms, corresponding to each contribution. Depending on the distance z to the interface, one of them may dominate. Concerning spatial coherence, surface waves such as surface-plasmon polaritons or surface-phonon polaritons yield long-range spatial coherence on a scale of the surface wave propagation length which may be much larger than the wavelength when absorption is small. These are the dominant contributions at a distance from the surface of the order of a wavelength in vacuum. On the contrary, skin-layer currents and quasi-static fields dominate at a distance much shorter than the wavelength. They lead to a much shorter spatial coherence length that only depends on the distance to the interface [11]. Note, however, that this conclusion is based on the assumption of a local medium. The ultimately limiting scale is thus given by the electron screening length or the electron Fermi wavelength, whatever is larger.

The interested reader is refered to [11] for further details. In what follows, we will give a heuristic description of the mechanism that transforms a random source with uncorrelated currents into a spatially coherent source. Let us consider the contribution of free electrons in a metal for the sake of clarity. The currents produced by the electrons have two components: a random component due to the random thermal motion and an induced component due to the fields in the medium. The fluctuation-dissipation theorem together with the assumption of a local medium implies that the fluctuating currents are delta correlated. The question that arises is how can we obtain fields which are correlated over tens of wavelengths from such incoherent currents?

It is known that two slits will produce interference in transmission when illuminated by the light coming from the sun. The reason is that the coherence of the field increases upon propagation. This is known as the Zernike–van Cittert theorem. It can be shown that the cohence length in the plane perpendicular to the beam is given by λ/θ, where θ is the angle subtended by the source. The reason is that a given point in the source illuminates both slits when they are far apart from the source plane. This creates a correlation for the fields at both slits. On the contrary, if the slit is close to the source, it is mainly illuminated by points of the sources lying just beneath it. Therefore, the fields at both slits are uncorrelated. This is exactly what happens when we consider the electric field very close to the interface. Then, the field is mainly due to the source elements which lie just beneath the observation point, because the electrostatic components varying like $1/r^3$ dominate. This is why we observe a very short coherence length in Fig. 9 for tungsten. The typical coherence length depends only on the distance between the observation point and the source plane.

What happens now if there is a surface wave? Any point source can excite a surface wave. Since a surface wave is a delocalized mode of the system, it builds up over the interface with a spatial extension given by its decay length. Therefore, each volume element radiating light can excite a surface wave that spreads over tens of wavelengths along the interface. There is an analogy with a piano chord. The source is a hammer that strikes the chord at a single point. Yet, the mode of the chord is excited and vibration takes place along the full length of the chord. This is an example of a delocalized mode. For a thermal light source, the source current due to the random thermal motion of the charges is random and uncorrelated. However, the induced currents due to the excitation of the surface wave are delocalized and produce a large coherence length.

5 A Spatially Coherent Thermal Source

We have just discussed the spatial coherence of a thermal planar source. In this section we describe an experimental result [12] that proves that thermal sources can be partially spatially coherent. The basic consequence of the transverse spatial coherence for a laser is that it propagates with a narrow angular divergence approximately given by the ratio λ/w, where w is the beam waist. By contrast, as mentionned in the introduction, a thermal source such as a light bulb is a quasi-lambertian source. This is related to the lack of coherence of the source. Consider two different points of the source; if they are uncorrelated, the fields that they emit cannot interfere. Since each point has a quasi-lambertian angular emission pattern, the overall emission angular pattern is also quasi-lambertian.

We have seen that the field may be coherent along the interface. However, since this coherence is only due to surface waves, it cannot be detected in the far field. To transfer the near-field coherence into the far field, we can rule a grating on the interface. The surface waves are then coupled with propagating waves. Since they propagate along distances of the order of $l \sim 10$–20λ, the field scattered by each groove of the grating can interfere, producing maxima of emission in well-defined directions with an angular width approximately given by λ/l. The particular properties of such sources were first reported by *Hesketh* et al. [20] and later by *Kreiter* et al. [21], although the role of surface waves and coherence was not fully understood at that time. Emission by surface waves using coupling by a prism has also been widely studied by *Zhizhin* et al. [22].

In order to produce such a source, it is essential to properly design the grating so that the coupling efficiency of the grating is as high as possible. To this end, we have used numerical simulations as described in [23,24] in order to find the optimum parameters. The material chosen was SiC, because this material may support a surface wave in the spectral range close to 10 µm, which is roughly the peak wavelength of the Planck spectrum at ambient temperature. The grating was then fabricated using standard techniques. An image taken with an atomic force microscope is seen in Fig. 11. The angular emission measurements at 773 K are shown in Fig. 12. The most striking feature is the fact that the emission pattern looks like an antenna emission pattern. This is a signature of the partial spatial coherence of the source, as discussed in [12]. An indirect measurement of the emissivity ε can be done

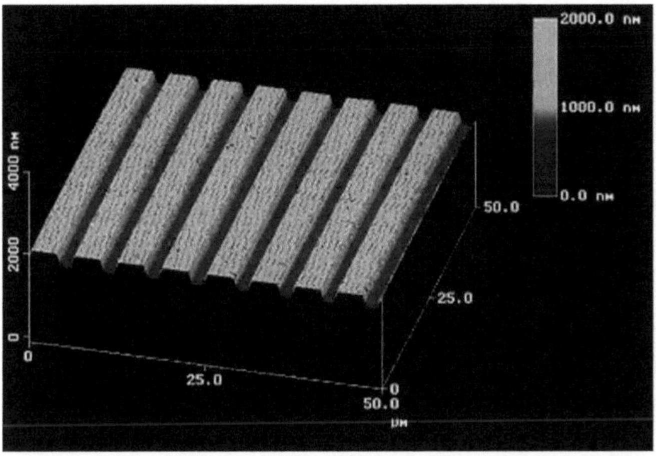

Fig. 11. Atomic force microscope image of the grating. The period is 6.25 µm and the height is 0.284 µm. The parameters were optimized so that the emissivity is *l* for a wavelength 11.36 µm. Reproduced from [12]

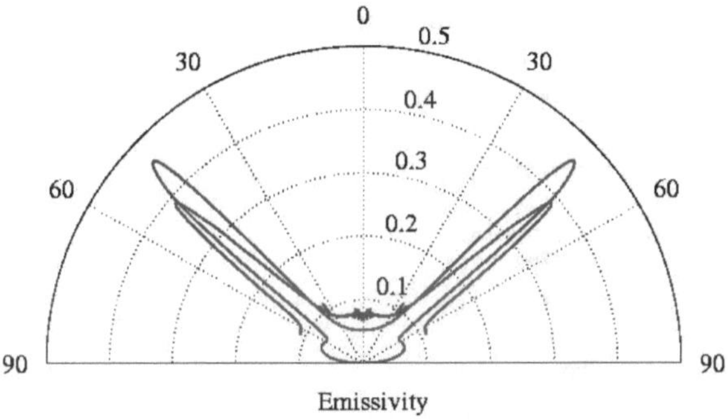

Fig. 12. Polar plot of the emissivity of the grating at a wavelength 11.36 µm and temperature 773 K. *Red*: experimental data; *green*: theoretical calculation. The theoretical result has been convoluted with an angular window to account for the experimental angular resolution. The disagreement is due to the fact that the calculations use the values of the optical constants at ambient temperature. It is seen that most of the light is emitted in a narrow angular cone. Reproduced from [12]

by measuring the reflectivity R of the sample. Indeed, Kirchhoff's [25] law states that $R = 1 - \varepsilon$.

We were able to measure the dispersion relation of the surface-phonon polariton by studying the spectral reflectivity of the grating. The result is shown in Fig. 13. By looking at the lower branch of the dispersion relation it is seen that a particular frequency is associated with a particular component of the wave vector parallel to the interface. The wave vector is related to the emission direction in the far field by the simple relation $k_{||} = (\omega/c) \sin\theta$. It is thus seen that the spectrum depends on the angle of observation. This is

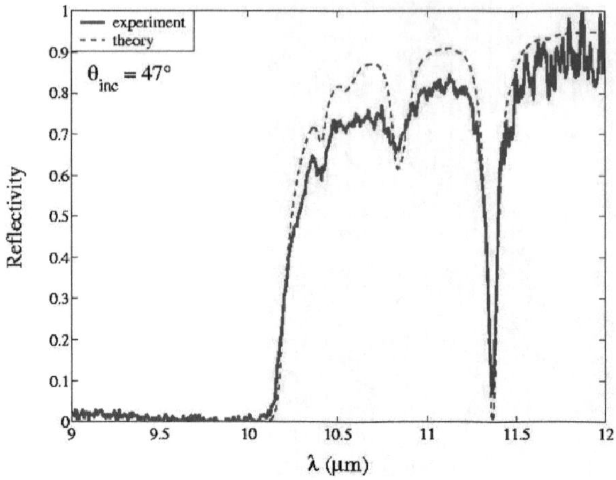

Fig. 13. Reflectivity of p-polarized light by the grating as a function of the wavelength. The dip observed at $\lambda = 11.36\,\mu\text{m}$ coincides with the emission peak observed in Fig. 12. The figure shows clearly a quasi-total absorption of the incident light due to the resonant excitation of a surface-phonon polariton. A flat surface would have a reflectivity on the order of 0.94. Note the very good agreement between experiment and theory. This agreement shows that the disagreement in the emission measurements is due to the change of the index with temperature

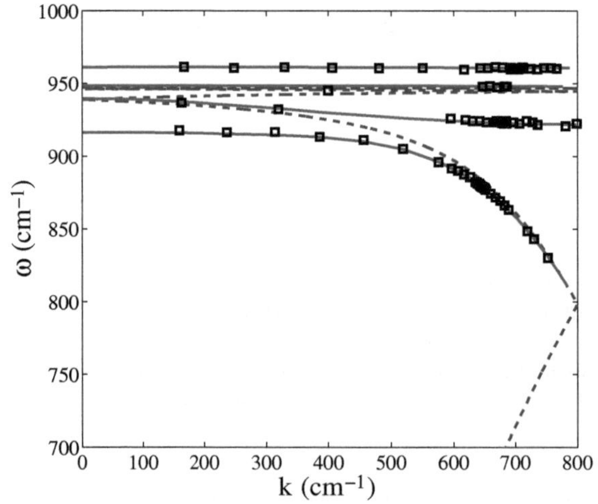

Fig. 14. Dispersion relation of the surface-phonon polariton on the grating. *Green curve*: theoretical dispersion relation. *Red curve*: theoretical dispersion relation for the flat surface. Data points: experimental measurements. The dispersion relation is constructed from reflectivity measurements. A spectrum is taken of the specular reflectivity of the grating at a fixed angle θ using a Fourier Transformation Infra-Red spectrometer. The frequency points of each minimum of the reflectivity and the value $(2\pi/\lambda)\sin\theta$ yield the coordinates of a point of the dispersion relation. It is seen that there is excellent agreement between the experiment and the theory

standard behavior for any coherent source such as an antenna. For random sources, it is far from trivial. It is a manifestation of the interplay between the propagation and correlation of the field known as the Wolf effect [26]. It has been observed for secondary partially coherent sources so far. The grating has this property. Indeed, by making a measurement of emission in the far field, one selects a particular angle and therefore a particular wave vector. The dispersion relation of the surface waves that produce the coherence shows that for a given wave vector, there will be peak emissions at some frequencies. If we now consider higher frequencies, it is seen that the dispersion relation does almost not depend on the wave vector. This means that the emission is almost isotropic, with a large peak at some particular frequencies where the dispersion relation has a flat asymptote. This can be used to design infrared light emitters.

6 Conclusions

In this paper, we have reviewed recent results concerning the near-field properties of thermal sources of light. The existence of surface waves produces a very intense peak of the local density of states close to the interface for some particular frequencies. We have highlighted several new properties of thermal sources in the near field by exploring the implications of this large density of states.

We have shown that thermal sources may produce quasi-monochromatic near fields. In light of this result, the possibilities of performing near-field solid-state spectroscopy and of designing near-field infrared sources have been discussed. The implication for photovoltaic applications has also been discussed.

The problem of radiative transfer between two thermal sources held at subwavelength distance has been studied. We have shown that the radiative flux may be enhanced by several orders of magnitude due to the excitation of resonant surface waves, and that it may occur at particular frequencies.

Finally, we have studied the spatial coherence of thermal sources and the substantial influence of the near field. We have shown that surface waves may induce long-range spatial correlations on a scale much larger than the wavelength. Sources that emit light within narrow angular lobes in the far field have been demonstrated. Conversely, quasi-static contributions, as well as skin-layer currents, induce correlations on small length scales (as far as a macroscopic and local description of the materials is correct).

With the recent development of local (optical and thermal) probe microscopy and the advent of nanotechnology, it is necessary to revisit the emission of light by plane surfaces. The results presented in this paper show the crucial role of surface waves in this respect.

References

1. L. Mandel, E. Wolf: *Optical Coherence and Quantum Optics* (Cambridge University Press, Cambridge 1995) 163, 165, 174, 175
2. A. V. Shchegrov, K. Joulain, R. Carminati, J.-J. Greffet: Phys. Rev. Lett. **85**, 1548 (2000) 163, 167, 169
3. C. L. Tien, G. Chen: J. Heat Transfer **114**, 799 (1994) 164, 170
4. A. R. Abramson, C. L. Tien: Microscale Thermophys. Engin. **3**, 229 (1999) 164
5. C. C. Williams, H. K. Wickramasinghe: Appl. Phys. Lett. **49**, 1587 (1986) 164, 170
6. E. G. Cravalho, C. L. Tien, R. P. Caren: J. Heat Transfer **89**, 351 (1967) 164, 171
7. D. Polder, M. Van Hove: Phys. Rev. B **4**, 3303 (1971) 164, 171, 172
8. J. J. Loomis, H. J. Maris: Phys. Rev. B **50**, 18517 (1994) 164, 171, 172
9. J.-P. Mulet, K. Joulain, R. Carminati, J.-J. Greffet: Appl. Phys. Lett. **78**, 2931 (2001) 164
10. R. Carminati, J.-J. Greffet: Phys. Rev. Lett. **82**, 1660 (1999) 164, 165, 171, 175, 176
11. C. Henkel, K. Joulain, R. Carminati, J.-J. Greffet: Opt. Commun. **186**, 57 (2000) 164, 167, 176, 177
12. J. J. Greffet, R. Carminati, K. Joulain, J. P. Mulet, S. Mainguy, Y. Chen: Coherent emission of light by thermal sources, Nature **416**, 61 (2002), 164, 178, 179
13. S. M. Rytov, Yu. A. Kravtsov, V. I. Tatarskii: *Principles of Statistical Radiophysics* **3** (Springer Verlag, Berlin, Heidelberg 1989) Chap. 3 165, 167, 171
14. L. D. Landau, E. M. Lifshitz: *Electrodynamics of Continuous Media* (Pergamon, Oxford 1960) 168
15. E. W. Palik: *Handbook of Optical Constants of Solids* (Academic Press, San Diego 1985) 165, 169, 175
16. K. Joulain, J.-P. Mulet, R. Carminati, J.-J. Greffet, A. V. Shchegrov: Proc. Heat Transfer and Transport Phenomena in Microsystems Conf., Banff, Canada, 2000 169
17. V. M. Agranovich, D. L. Mills (Eds.): *Surface Polaritons* (North-Holland, Amsterdam 1982) 166, 175
18. J. A. Porto, R. Carminati, J.-J. Greffet: J. Appl. Phys. **88**, 4845 (2000) 169
19. J. Xu: *Heat Transfer Between Two Metallic Surfaces at Small Distance*, Dissertation, University of Konstanz, Germany (1993) 171, 172
20. P. J. Hesketh, J. N. Zemel, B. Gebhart: Nature (London) **324**, 549 (1986) 178
21. M. Kreiter, J. Oster, R. Sambles, S. Herminghaus, S. Mittler-Neher, W. Knoll: Thermally induced emission of light from a metallic diffraction grating, mediated by surface plasmons, Opt. Commun. **168**, 117–122 (1999) 178
22. G. N. Zhizhin, E. A. Vinogradov, M. A. Moskalova, V. A. Yakovlev: Appl. Spectrosc. Rev. **18**, 171 (1982) 178
23. A. Sentenac, J. J. Greffet: J. Opt. Soc. Am. A **9**, 996 (1992) 178
24. J. Le Gall, M. Olivier, J.-J. Greffet: Phys. Rev. B **55**, 10105 (1997) 178
25. J. J. Greffet, M. Nieto-Vesperinas: J. Opt. Soc. Am. A **10**, 2735 (1998) 179
26. E. Wolf: Phys. Rev. Lett. **56**, 1370 (1986) 181

Plasmon Resonances in Nanowires with a Non-regular Cross-Section

Olivier J. F. Martin

Electromagnetic Fields and Microwave Electronics Laboratory,
ETZ-G46, ETH-Zentrum, 8092 Zürich, Switzerland
martin@ifh.ee.ethz.ch

Abstract. We investigate numerically the spectrum of plasmon resonances for metallic nanowires with a non–regular cross-section in the 20–50 nm range. After briefly recalling the physical properties of metals at optical frequencies, we point out the intrinsic difficulties in the computation of the plasmon resonances for nanoparticles with a non–regular shape. We then consider the resonance spectra corresponding to nanowires whose cross-sections form different simplexes. The number of resonances strongly increases when the section symmetry decreases: A cylindrical wire exhibits one resonance, whereas we observe more than 5 distinct resonances for a triangular particle. The spectral range covered by these different resonances becomes very large, giving to the particle specific distinct colors. At the resonance, dramatic field enhancement is observed at the vicinity of non–regular particles, where the field amplitude can reach several hundred times that of the illumination field. This near–field enhancement corresponds to surface enhanced Raman scattering (SERS) enhancement locally in excess of 10^{12}. The distance dependence of this enhancement is investigated and we show that it depends on the plasmon resonance excited in the particle, i.e. on the illumination wavelength. The average Raman enhancement for molecules distributed on the entire particle surface is also computed and discussed in the context of experiments in which large numbers of molecules are used. Finally we discuss the influence of the model permittivity which enters the calculation, as well as the resonances shift and broadening produced by a water background.

1 Introduction

The interaction of light with small metal particles has been of great interest for many centuries. Medieval artisans, for example, made use of metal colloidal particles in the production of certain types of stained glass. By tuning the particle size and composition, they were able to produce glasses with specific colors, which were the signatures of the plasmon resonances excited in the particles. Since a plasmon resonance does not wear out, we can still enjoy these shiny colors today. The fact that there is a connection between the scattered spectrum of a nanoparticle and its physical properties was established long ago. For instance, Faraday noted "that a mere variation in the size of its particles gave rise to a variety of colors" [1]. This fact can easily be observed in a solution of colloidal gold or silver under the optical

J. Tominaga and D. P. Tsai (Eds.): Optical Nanotechnologies,
Topics Appl. Phys. **88**, 183–210 (2003)

microscope: although all particles have a similar composition, they shine with several different colors.

The relation between the shape and the spectrum was recently highlighted in a beautiful experiment by *Schatz* and co-workers [2]. The transformation of the plasmon-resonance spectrum was correlated to the changes of the form of the silver nanoparticles as they evolved from a spherical shape into a prism-like shape. A similar study was presented by *Mock* et al., who systematically investigated the effect of size and shape on the spectral response of individual silver nanoparticles [3].

Beyond creating nice color effects, the current interest in colloidal metals is driven by two major topics. On the one hand, there is the phenomenon of surface-enhanced Raman scattering (SERS), wherein the Raman signal of ensembles of molecules adsorbed on rough metal surfaces can be enhanced by a factor of 10^7 [4,5,6,7]. In recent experiments, Raman enhancements of up to 10^{12} were even reported for single molecules located on so-called "hot spots" [8,9,10]. The adsorption of the molecule on the metal can participate in the Raman enhancement [11,12]. However, it is believed that the excitation of plasmons in the metallic nanoparticles creates greatly enhanced local electromagnetic fields that contribute the major component of the SERS effect.

On the other hand, a new field of nano-photonics is currently coming to life, where plasmon-resonant nanoparticles are used to guide and switch light at the nanoscale. New classes of waveguides, which utilize plasmon resonances as means of transporting electromagnetic energy have been demonstrated [13,14,15]. The utilization of plasmon-resonant nanoparticles can also dramatically reduce the spot size, thereby increasing the data storage density for next-generation optical data storage disks [16,17]. Finally, the utilization of plasmon-resonant particles as switchings elements opens exciting perspectives for new active devices [18]. These different active and passive components may pave the way for the all-optical dream: a network where information transmission and switching occur entirely at the optical level.

In order to successfully implement plasmon resonances both in chemistry and in optics, it is mandatory to further our understanding of the underlying physics for nanoparticles with an arbitrary shape. This knowledge should allow us to realize plasmon-resonant particles with tailored properties for specific applications. For example, by tuning the plasmon resonance frequency one could fabricate particles adapted to specific chemical compounds or produce particles with several well-defined resonances, covering a given frequency range. Also, the near-field distribution in the vicinity of the particle could be tuned and specific field distributions which would provide the strongest enhancement at the location of the active molecular site could be developed. Finally, the plasmon resonance linewidth, which is related to the plasmon lifetime – an important parameter for transient and nonlinear phenomena – could be tuned by changing the particle properties.

It should be noted that considerable progress has been made in recent years in the fabrication of metallic nanostructures in a controlled manner, including features in the 10–50 nm range [19,20,21,22,23,24,25,26,27,28]. Metallic particles with a variety of shapes and dimensions are now readily available for experiments. A thorough understanding of the detailed local fields associated with plasmon-resonant particles is therefore warranted, to enhance the design and optimization of applications based on these particles.

The accurate solution of the local fields of plasmon-resonant particles of arbitrary shape remains, however, a theoretical challenge. Analytical solutions for the fields are known only for particles with a very simple shape, like that of a sphere or an ellipsoid [29,30,31,32,33,34,35,36,37], or spherical shells and periodic cylinder gratings [38,39,40,41,42,43,44]. While electrostatic methods can provide some level of insight [45], complete electromagnetic solutions are needed to obtain accurate results. Many groups have developed methods of solving Maxwell's equations to investigate non-regular plasmon-resonant particles in the 100–200 nm size range [46,47,48]; however, although particles of this size provide a large scattering cross-section (SCS) at the plasmon resonance, the resonances are very broad and the field enhancement in the vicinity of the particles is relatively small.

Over the last few years we have elaborated a numerical approach for the solution of the fields associated with plasmon-resonant nanoparticles of arbitrary two-dimensional geometry, which leads to highly accurate, converged solutions, even for particles having extremely large local enhancement and field variation [49]. With this technique we were able to study some aspects of the plasmon resonances in nanowires with a non-regular cross-section in the sub-100 nm size range [50,51,52,53]. Several on-line publications have also illustrated with movies the dynamics of specific effects [54,55]. In the present chapter, we shall summarize these results and draw some conclusions for the utilization of plasmon engineering in the nanosciences. We will only consider the case of individual nanowires and not address that of interacting nanowires, where interesting effects can also occur [56,57].

Since our work is theoretical, it appears important to first sketch in Sect. 2 the difficulties associated with the simulation of plasmon-resonant nanoparticles. In Sect. 3 we investigate the relationship between a nanowire cross-section and its plasmon resonance spectrum. An extremely useful tool for understanding the characteristics of each plasmon resonance is its associated polarization charge distribution, as discussed in Sect. 4. Some implications for SERS are discussed in Sect. 5, while the limitation of our Maxwellian approach is addressed in Sect. 6. Conclusions and outlook are given in Sect. 7.

2 Model

2.1 Metals at Optical Frequencies

The electromagnetic scattering of metal nanoparticles can be described by solving Maxwell's equations [58,59]. In this model, the complete description of the material properties of the metal is encompassed in the dispersion relation, which gives the complex permittivity $\varepsilon(\omega)$ as a function of the frequency (or wavelength). The dispersion relation measured by *Johnson* and *Christy* for silver is shown in Fig. 1 [60]. We shall use these values for our calculations.

Near the plasma frequency ω_p of the metal, the dispersion relation is governed by the interaction between light and the conduction electron gas, or expressed with their quantum counterparts by the photon–plasmon interaction. The combined exciton is often referred to as plasmon–polariton. For certain metals such as silver, copper and gold, ω_p is in the visible frequency range. At specific negative permittivity values, plasmon resonances will be excited in these small particles. These specific permittivity values strongly depend on the particle size and shape, since the boundary conditions imposed by Maxwell's equations determine whether such a particle resonance can build up. These resonances are often referred to as the surface modes of the particle [61].

The plasmon resonances are analytically known only for simple geometries, such as a sphere or a cylinder [58]. In a very small sphere, for instance, one single resonance can be excited, when $\varepsilon = -2$ (Ag: $\lambda \simeq 355\,\mathrm{nm}$, Au: $\lambda \simeq 490\,\mathrm{nm}$), whereas a cylinder is in resonance when $\varepsilon = -1$ (Ag: $\lambda \simeq 337\,\mathrm{nm}$, Au: $\lambda \simeq 253\,\mathrm{nm}$). With increasing particle size, these resonances are red-shifted and broadened, and additional higher-order resonances can be excited [58].

More than one single resonance can be excited in a non-regular structure, irrespective of its size. A cylinder with elliptical cross-section, for instance, exhibits two resonances, corresponding to the illumination directions along and normal to its major axis (Fig. 2a). Recently we demonstrated that triangular nanoparticles have several resonances for each illumination direction [50].

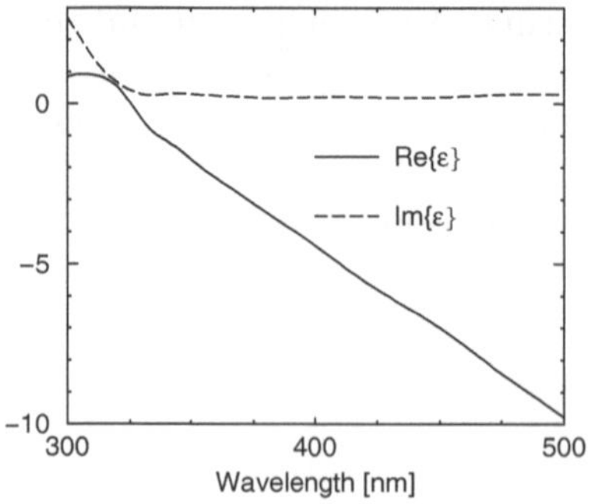

Fig. 1. Dispersion relation for silver used in our calculations. The real ε' and the imaginary ε'' parts of the permittivity are interpolated from the experimental data of *Johnson* and *Christy* [60]

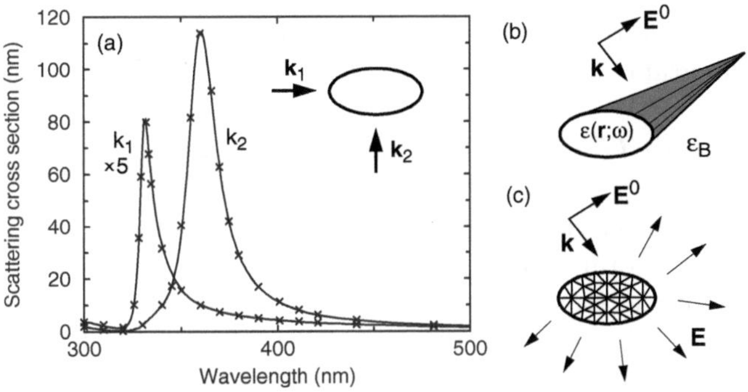

Fig. 2. (**a**) Scattering cross-section for an elliptical silver cylinder (40 nm × 20 nm section) computed with finite elements (*solid line*) and Mie theory (*cross*). Two orthogonal propagation directions $\boldsymbol{k}_1$ and $\boldsymbol{k}_2$ for the incident field, corresponding to two different resonances, are considered. The simulation of (**b**), an infinitely long nanowire, can be limited to its cross-section, (**c**). In our computational approach, the cross-section is discretized using triangular finite elements

Whether these different modes of a nanoparticle of arbitrary shape can be resolved, strongly depends on the material absorption (the imaginary part of ε): large absorption broadens the resonances, and can result in a nearly featureless band. Silver, compared with other metals that have their plasma frequency in the optical range, has a comparatively low absorption, and thus narrower resonances.

One may wonder whether the classical description of the material that we use, based solely on Maxwell's theory and a local dispersion relation, is appropriate for the small structures investigated here. Actually, it has been experimentally shown that this macroscopic approach is adequate for particle dimensions as small as 5 nm [37,59,62,63,64]. Quantum effects must only be taken into account for smaller particles, using for example a jellium or quantum-chemical model [59,65,66,67]. For particles in the 5–20 nm range, the dispersion relation depends noticeably on the particle geometry, since the electron mean free path decreases as electron scattering at the surface becomes more important [37,62,63,64]. However, for the silver particles investigated here, only the imaginary part of the permittivity increases slightly, and the bulk values of permittivity still represent a good approximation. This point will be addressed in greater detail in Sect. 6.

2.2 Scattering Problem

As established in the previous section, Maxwell's equations are well suited to the study of the plasmon resonances of silver particles in the 20–50 nm range, and we shall use the experimentally obtained permittivity values from *Johnson* and *Christy* [60]. For non-magnetic media, Maxwell's equations reduce in the frequency domain to the vectorial wave equation, which is formally solved by the volume integral equation:

$$\boldsymbol{E}(\boldsymbol{r};\omega) = \boldsymbol{E}^0(\boldsymbol{r};\omega) + \int_V \mathrm{d}\boldsymbol{r}' \boldsymbol{G}^{\mathrm{B}}(\boldsymbol{r},\boldsymbol{r}';\omega) \cdot k_0^2 \left[\varepsilon(\boldsymbol{r}';\omega) - \varepsilon_{\mathrm{B}}\right] \boldsymbol{E}(\boldsymbol{r}';\omega) . \quad (1)$$

Here, $\boldsymbol{E}^0(\boldsymbol{r};\omega)$ is the incident electric field with vacuum wavenumber $k_0 = \omega/c$, $\boldsymbol{E}(\boldsymbol{r};\omega)$ is the unknown total scattered field, $\varepsilon(\boldsymbol{r};\omega)$ is the particle permittivity and ε_{B} is that of the background. The dyadic $\boldsymbol{G}^{\mathrm{B}}(\boldsymbol{r},\boldsymbol{r}';\omega)$ is the Green's tensor associated with the infinite homogeneous background [68]. This formalism can also be used when the background is a surface or a stratified medium [69,70]. Integrating the scattered field over all scattering angles provides the scattering cross-section (SCS) associated with the particle [58]. At the plasmon resonance the SCS shows distinct peaks, as the amount of scattered light increases abruptly. By repeating this calculation for different illumination frequencies ω with the corresponding permittivities $\varepsilon(\boldsymbol{r};\omega)$ one obtains the spectral response of the particle, as illustrated in Fig. 2a.

This so-called frequency-domain approach allows us, therefore, to take into account an arbitrary dispersion relation for the particle. The drawback is that the calculations must be repeated for each frequency where the particle response is required. An alternative approach, which provides the response of the particle over a certain frequency range in a single calculation, consists in using a time-domain technique such as the finite-difference time-domain technique [71]. In this case the dispersion relation must be included directly in the algorithm, which is only possible for the most simplistic dispersion models.

For infinitely long particles (cylinder, nanowire) under arbitrary illumination $\boldsymbol{E}^0$ (Fig. 2b), the scattering problem can be reduced to one section of the particle (Fig. 2c), [68]). We shall only consider illumination where the particles are illuminated in the plane of the figures with the electric field in the same plane (transverse-electric (TE) polarization). For transverse-magnetic (TM) polarization, where the incident magnetic field is in the plane, plasmons cannot be excited.

Equation (1) forms the basis for several different computational techniques for scattering calculations. The most widely used technique, the coupled dipole approximation (CDA), also known as the discrete dipole approximation (DDA), corresponds to solving (1) with finite differences [46,47,48,68] [72,73,74]. Unfortunately, experience shows that for high-permittivity scatterers, or small scatterers (less than 100 nm size) that support plasmon resonances, this technique does not converge properly [10,47,75]. This means that increasing the number of discretization elements does not always improve the accuracy of the results. Worse, there can be cases where a thin mesh produces less results of the accurate than a coarse one [75]. This implies that subtle features of the resonance spectrum cannot be resolved and it is not possible to obtain a quantitative description of the field distribution in the close vicinity of the particle. However, the CDA/DDA can provide useful insight into the plasmon resonances for larger particles.

Recently, we became convinced that the flaw of the CDA/DDA is related to the singular behavior of the Green's tensor in (1) for $\boldsymbol{r} \to \boldsymbol{r}'$. It is difficult to handle this singularity properly using a finite differences discretization because it assumes a constant field over each mesh. Refining the mesh requires the computation of the Green's tensor for neighboring points $\boldsymbol{r}$ and $\boldsymbol{r}'$ that become closer and closer, which leads to exploding values for $\boldsymbol{G}^{\mathrm{B}}(\boldsymbol{r}, \boldsymbol{r}'; \omega)$. To overcome this difficulty, we developed a new technique based on finite elements [49]. The arbitrary particle section is discretized using triangles, and the unknown field $\boldsymbol{E}(\boldsymbol{r}; \omega)$ is expanded into a sum of basis functions defined on each triangle (Fig. 2c). A Galerkin test procedure is then used to obtain a system of algebraic equations for the unknown field. We refer the reader to [49], where this technique is described in detail. Let us only mention that the use of finite elements allows us to handle exactly the singularity of the Green's tensor using generalized functions.

Since the spectrum of resonances for non-regularly shaped particles is not known, it is very important to first assess the suitability of this technique for investigating plasmon resonances. To do so, we resort to geometries where a reference solution exists. Figure 2a illustrates the accuracy of this approach for an elliptical particle. The entire spectrum agrees perfectly with that obtained semi-analytically using Mie's theory [58] and the location, magnitude and width of both plasmon resonances are reproduced. To quantify the accuracy of this technique, we show in Fig. 3 the error on the SCS as a function of the number of discretization elements. This error is defined as the square of the norm of the difference between the scattered field obtained with Mie's theory and that obtained with finite elements. Accurate results are already obtained with few discretization elements and, contrary to the CDA/DDA, it is always possible to increase the results accuracy by refining the mesh (Fig. 3). A key component in achieving these highly accurate results is the new regularization scheme that was developed to extract the singularity of the

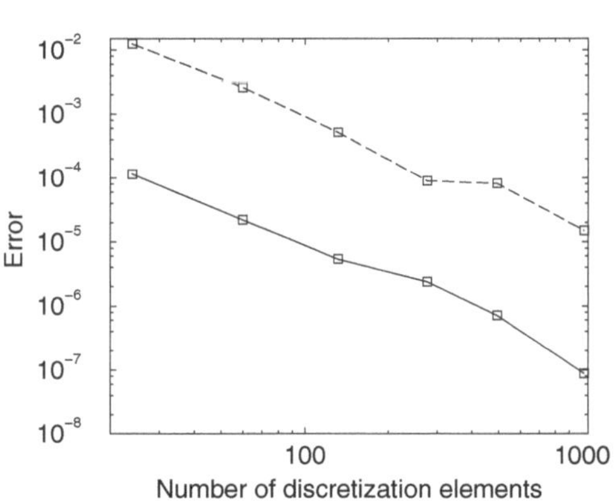

Fig. 3. Convergence of our finite-element technique. The absolute error as a function of the number of discretization elements is reported. Two cylinders with permittivity $\varepsilon = 16.64 + 0.23i$ (silicon, *dashed line*) or $\varepsilon = 4.0$ (*solid line*) are used as test objects. Their diameter is 100 nm and the incident wavelength $\lambda = 546$ nm; TE polarization

Green's tensor. This regularization scheme was also extended to neighboring elements, which further improves the overall accuracy and convergence [49].

The non-regular cross-sections investigated in this chapter were discretized with two to three thousand triangular elements. We verified that this discretization number was large enough so that the near-field distributions did not depend on it. We will consider nanowires with a polygonal section; dealing with sharp corners introduces additional numerical difficulties, since the field becomes singular at short distances from an infinitely sharp, perfectly conducting corner [76]. However, the sharpness of a real particle is limited by surface and boundary energies; therefore, we have rounded off each corner by 0.25 nm, providing a more realistic model and removing the numerical instabilities. It is demonstrated in Sect. III. D of [52] that this minute corner smoothing does not at all influence the plasmon resonance SCS or near-field distribution.

3 Relation Between Shape and Resonance Spectrum

Having assessed in the previous section the suitability of our approach for the simulation of plasmon resonances in metallic nanowires, we consider in this section wires with a non-regular shape and investigate the relationship between that shape and the spectrum of resonances supported by the particle.

Figure 4 shows the scattering cross-section (SCS) from nanowires having cross-sections corresponding to that of a circle, a hexagon, a pentagon, a square and a triangle. The illumination direction is along one of the particle's symmetry axes. Two different sets of simplexes are considered: in Fig. 4a all particles have the same area as a 20 nm circle, whereas in Fig. 4b they have that of a 50 nm circle. Since all particles in the same figure part have the same area, their SCS should be comparable.

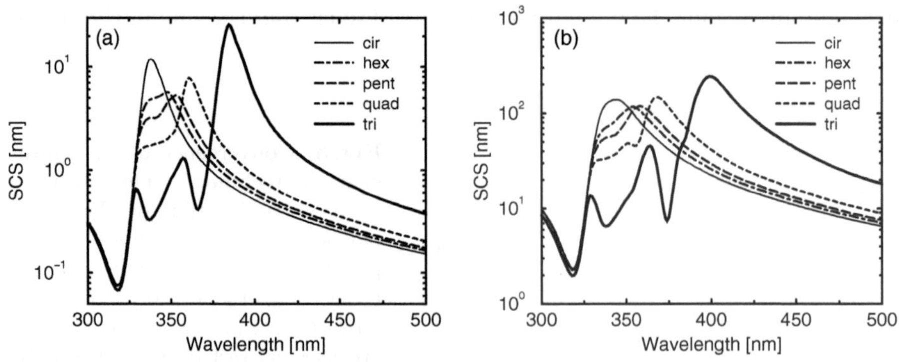

Fig. 4. SCS for (**a**) 20 nm and (**b**) 50 nm simplexes of varying symmetry (circle, hexagon, pentagon, square and triangle)

Let us first consider the 20 nm simplexes (Fig. 4a). For the circle we recover the well-known result of a single resonance, at $\lambda = 338$ nm (corresponding to $\varepsilon = -1.07 + 0.29i$). The full-width half-maximum (FWHM) of the resonance is about 25 nm. As can be seen in Fig. 4a, the structure of the SCS becomes more complex for less regular particles. The main resonance is red-shifted from $\lambda = 338$ nm (circle) to $\lambda = 350$ nm (hexagon), $\lambda = 357$ nm (pentagon), $\lambda = 361$ nm (square) and $\lambda = 385$ nm (triangle). An additional resonance appears for all the non-circular particles. Moreover, a third resonance can be observed for the triangle at $\lambda = 357$ nm. The influence of the illumination direction on the SCS is very small, due to the symmetry of the simplexes (not shown).

The large SCSs that occur at the plasmon resonances are associated with large field amplitudes at the vicinity of the particle. As for the SCS, the complexity of the field distribution increases when the symmetry of the particle decreases. For the circular particle, the amplitude is homogeneous inside the particle, about seven times the incident field, and rapidly decreases outside the particle. For the non-regular structures the field becomes strongly heterogeneous and the enhancement factor increases dramatically. At the main resonance, the field amplitude takes large values at the corners transverse to the incident wave vector. There the relative field amplitude for the 20 nm simplexes (Fig. 4a) exceeds 20 for the hexagon and the pentagon, 70 for the square and 150 for the triangle.

Let us now consider the larger 50 nm simplexes (Fig. 4b). Two main differences are observed with respect to the SCS of the smaller simplexes (Fig. 4a). First, for all shapes the main resonance is red-shifted (e.g. for the triangle it is now at $\lambda = 399$ nm, compared to $\lambda = 385$ nm for the 20 nm triangle). Second, the resonance FWHM is roughly doubled (i.e. for the circle, it is now more than 50 nm, compared to 25 nm previously). We also see in Fig. 4b that an additional resonance can be resolved for the square at $\lambda = 351$ nm, whereas two additional resonances start to emerge for the triangle at approximately $\lambda = 350$ nm and $\lambda = 382$ nm.

The triangular particle appears to have a rather complex spectrum of plasmon resonances, much more complex than a square particle. To further investigate this effect, we report in Fig. 5 the SCS for nanowires with sections that evolve from a rectangular shape into a triangular one. The triangular section is such that it has the same area as a cylinder with a 20 nm diameter (Fig. 4a); the base is approximatively 17 nm and the height 34 nm. The rectangle has the same dimensions and the intermediate particles have a short side with length 0.75, 0.5, respectively, 0.25 times that of the triangle height. Let us note that the resonance spectrum covers a broader wavelength range for the right-angled triangle (Fig. 5) than for a similar equilateral triangle (Fig. 4a). The magnitude of the resonances is also larger in the case of the lowest-symmetry particle.

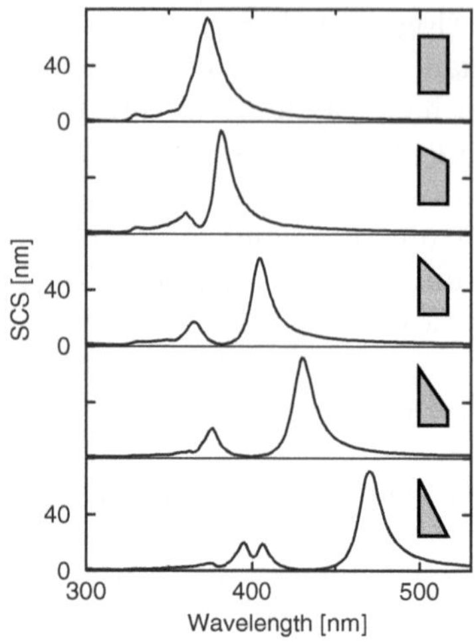

Fig. 5. SCS for nanowires with a cross-section evolving from a rectangular shape (*top*) to a triangular shape (*bottom*). The corresponding shape is shown in the inset; see the text for the dimensions. The illumination direction is along the lower left corner. The same vertical scale is used for the five curves

The FWHM for the main resonances in Fig. 5 is in the order of 15 nm and, surprisingly, does not vary much with the particle shape. It is, however, difficult to correlate the FWHM with the particle geometry only, as the main resonance occurs at a different wavelength, i.e. for a different permittivity (see Fig. 1). For example, the particle with the least-cut corner (Fig. 5, second from the top) has the narrowest resonance, FWHM=14 nm, but this resonance takes place at the wavelength where the absorption is minimum ($\varepsilon'' = 0.18$), which narrows the resonance.

4 Polarization Charge Distributions

Corresponding to the complexity observed in the far field for the triangular particle, there is a very complicated near-field distribution which varies extremely rapidly when the illumination wavelength changes, as illustrated in the movies presented in [54].

To investigate these near-field effects in further details, let us consider a 10 nm base, 20 nm perpendicular right-angled triangular particle, whose SCS is shown in Fig. 6. The field distributions corresponding to the two resonances labeled in Fig. 6 are shown in Fig. 7a,b. In these figures, the field amplitude is color coded (unit incident amplitude), whereas the arrows give the direction of the electric field at a given time (the field is harmonic and oscillates over one period; movies illustrating this behavior are presented in [55]).

First, notice the dramatic field enhancement at close vicinity to the corner of the particle at the main resonance (Fig. 7a): the field reaches 200 times the amplitude of the incident field. This corresponds to an intensity enhancement of 40 000, or a Raman enhancement in excess of 10^9 for a molecule immersed

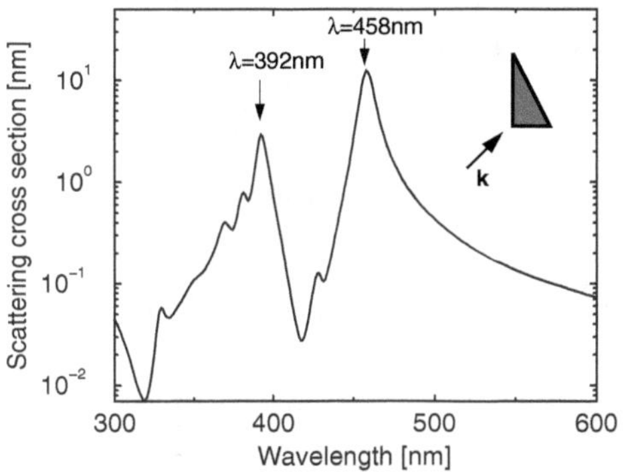

Fig. 6. SCS for a 10 nm base, 20 nm perpendicular right-angled triangle, as a function of the illumination wavelength

in this field distribution. This last point will be discussed in greater detail in Sect. 5.

The field distributions observed in Fig. 7a,b are typical of the different types of behavior observed in non-regular particles. For the main resonance ($\lambda = 458\,\text{nm}$) the field radiates radially from the "hot" particle's corner, where the field is extremely homogeneous. For the next resonance ($\lambda = 392\,\text{nm}$), although the enhancement is comparable to the main resonance, the topology of the field is very different. Now the field seems to turn around the particle's top corner; its spatial variation is very rapid with an amplitude going from over 100 to 0 within a few nanometers inside the particle tip. This extremely fast variation of the field distribution actually illustrates the difficulties related to the simulation of plasmon resonances in nanoparticles with an arbitrary shape: It requires a technique which is stable and accurate enough to handle these important field gradients.

To understand better the intrinsic properties of the different resonances illustrated in Fig. 7a,b, we show in Fig. 7c,d the corresponding polarization charge distribution, which is given simply by the divergence of the electric field [77]. Of course, the particle does not become charged by the effect of the external field: a similar amount of positive and negative charge is induced on its surface, so that the particle always remains neutral. Over one period, these charges move around the particle, as illustrated in the movies presented in [55]. The polarization charge distributions shown in Fig. 7c,d correspond to snapshots at an arbitrary moment in time. Half a period later, the reverse charge distribution is observed (with plus charges where there were negative ones, and vice-versa).

As for the field distribution, we notice in Fig. 7c,d that each resonance is associated with a different charge distribution. In the main resonance, for $\lambda = 458\,\text{nm}$, charges of a given sign build up at the sharp corner, while opposite charges are distributed on the entire circumference of the particle (Fig. 7c). This distribution oscillates over time, the sign of the accumulated charges on the sharp corner changing every half-period. This explains why the corresponding field distribution is radial, with the electric field either

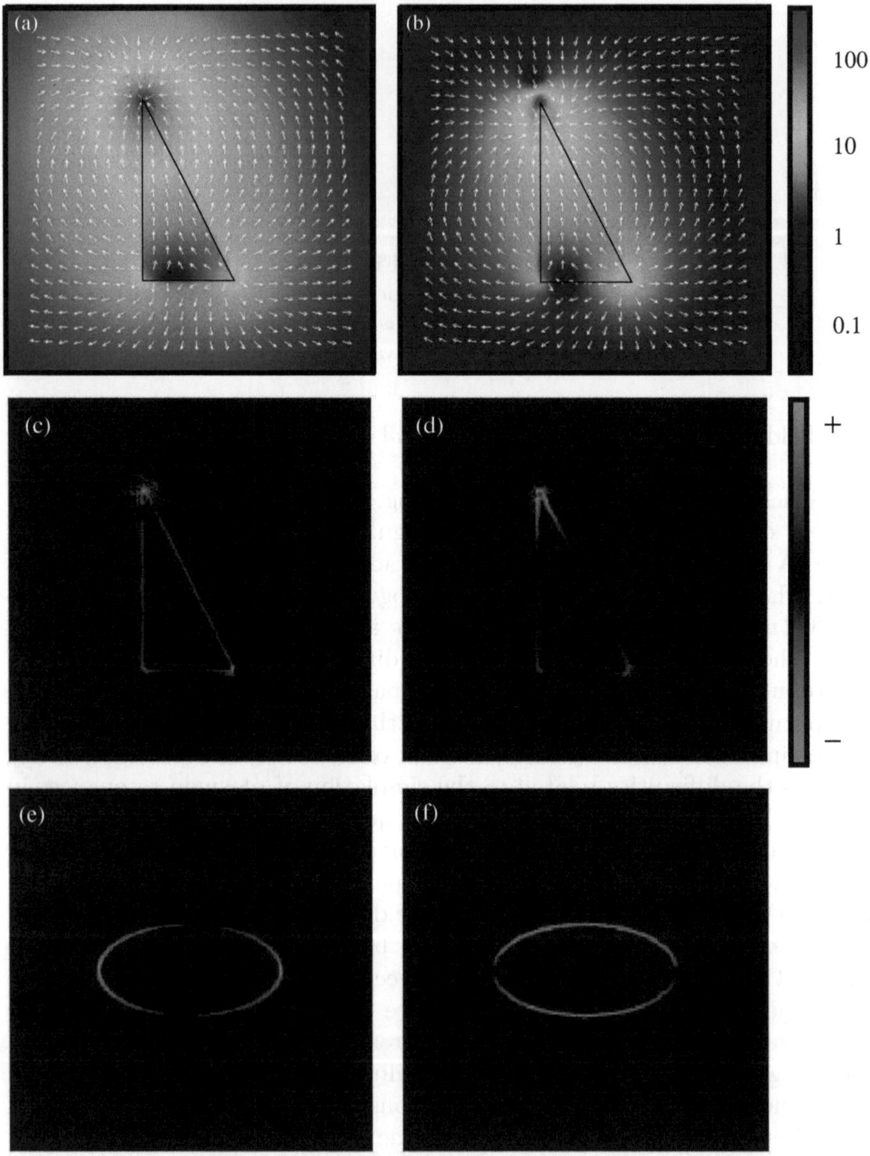

Fig. 7. (**a**),(**b**) Electric field amplitude distributions for the main [(**a**), $\lambda = 458\,\text{nm}$] and next-order [(**b**), $\lambda = 392\,\text{nm}$] resonances for a $10\,\text{nm} \times 20\,\text{nm}$ right-angled triangle (SCS in Fig. 6). The *arrows* show the orientation of the electric field. (**c**),(**d**) Polarization charge distributions associated with (**a**) and (**b**). (**e**),(**f**) Polarization charge distributions associated with the main [(**e**) $\lambda = 358\,\text{nm}$] and higher-order [(**f**) $\lambda = 331\,\text{nm}$] resonances for a $20\,\text{nm} \times 10\,\text{nm}$ elliptical particle

pointing towards or away from this corner, which is characteristic of a point source.

For the next resonance, $\lambda = 392\,\text{nm}$, both charge species accumulate simultaneously at the sharp corner: one species accumulates at the very tip, while the species of opposite sign is distributed along the adjacent sides (Fig. 7d). This dipolar-like charge distribution determines the field at the sharp corner, where the field appears to turn around the corner as it does at the vicinity of a dipole (Fig. 7b).

For higher-order modes, a more complex charge distribution is observed, as described in [55]. This charge distribution is extremely useful for understanding the topology of the electric field, both for individual and interacting nanoparticles. For example, when two particles interact together, new plasmon resonances can appear. The topology of the corresponding field can be easily related to the charge distribution on both particles [57,56].

To illustrate the topology of the charge distribution in a somewhat more intuitive case, Fig. 7e,f show the distributions associated with the two modes of an elliptical particle (SCS similar to that in Fig. 2a, although the particle is now smaller). For the fundamental mode ($\lambda = 358\,\text{nm}$, Fig. 7e) the electric field is parallel to the major axis, corresponding to polarization charges accumulating at both ends of this axis. For the higher-order mode ($\lambda = 331\,\text{nm}$, Fig. 7f), the polarization charges are split vertically, following the polarization direction of the incident field.

The polarization charge distributions are also helpful for understanding the spatial extension of the electric field away from the particle surface. In Fig. 7, for example, one notices that the spatial extension of the field is larger for the fundamental mode (Fig. 7a) than for the next-order mode, where the field decreases much faster (Fig. 7b). This is not surprising, since the former mode is related to a point-like charge distribution (3D spatial variation $\sim 1/r$), whereas the latter is associated with a dipolar charge distribution $\sim 1/r^3$). This difference of distance dependence as a function of excited resonance, i.e. as a function of illumination wavelength, could be demonstrated by the approach curves in scanning near-field optical microscopy (SNOM) experiments [51].

In that context of local probe microscopy, let us emphasize that the different plasmon resonances supported by a non-regular shape nanoparticle provide many different electromagnetic environments, while keeping the physical system unchanged. This could be used to suppress experimental artifacts by performing a series of measurements under different illumination wavelength and tracking the pure optical contrast [78].

Let us stress here that the large fields observed in Fig. 7a,b near the particle's corners are not produced by the lightning rod or tip effect [76]. The latter provides only a field amplification factor in the order of 5 to 10, even for very large permittivity values, as illustrated in Fig. 4c of [54].

5 Field Enhancement and SERS

In this section we discuss the near-field distributions associated with the plasmon resonances of the different simplexes, with emphasis on the local variation of the field amplitude around the particles. This variation plays a key role in the practical implementation of plasmon-resonant particles in different areas of nanoscience. For SERS in particular, the electromagnetic enhancement effectively experienced by a molecule depends on the location where the molecule is adsorbed on the metal, as well as the relative position of the Raman active site within the molecule. This becomes very important when the molecule is large and placed in a strongly inhomogeneous field. This was recently verified experimentally by *van Duyne* et al. , who inserted different numbers of non-active linker molecules between the adsorption site of a molecule and the Raman active site [79]. In this experiment, the Raman signal was highly dependent on the number of linker molecules, and therefore on the distance between the surface and the Raman active site, particularly for non-regular (tetrahedral) nanoparticles. Further, when a large number of molecules are adsorbed on the same nanoparticle, a spectral shift in the particle resonances can be observed [80].

Figure 8 shows the maximum enhancement obtained at a distance of 1 nm from the surface as a function of the wavelength. The amplitude enhancement is shown on the left-hand axis and the corresponding Raman enhancement on the right-hand; both 20 nm and 50 nm simplexes are investigated. For molecules excited far from an electronic absorption band, the intensity of the Raman scattered light is proportional to the fourth power of the local electric field amplitude where the molecule is immersed [7]. Plasmon-resonant particles provide a convenient method of enhancing the electromagnetic fields, and therefore are ideal SERS substrates. A second enhancement mechanism, the so-called chemical enhancement, related to the adsorption of the molecule on

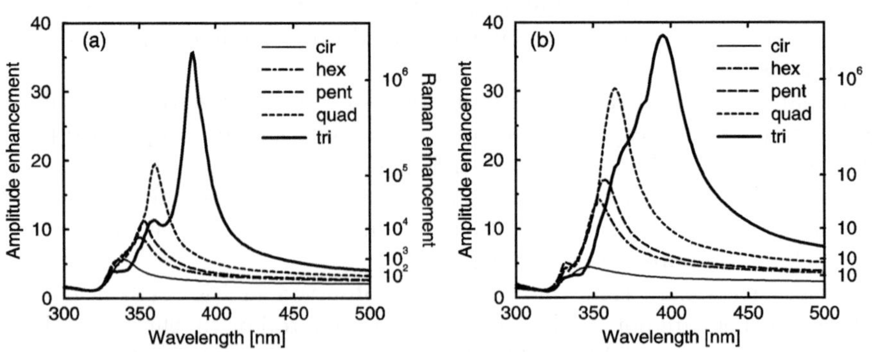

Fig. 8. Maximum amplitude enhancement for the (**a**) 20 nm and (**b**) 50 nm simplexes (SCS in Fig. 4). The right-hand scale shows the corresponding maximum Raman enhancement

the metal, can also contribute to SERS [11,12,81]. However, electromagnetic enhancement is believed to make the major contribution to SERS.

We observe that the field enhancement, which is a near-field quantity, is strongly correlated to the SCS, a far-field quantity (compare Figs. 8 and 4). The position of the main resonance is the same, but the resonance width is broader in the enhancement diagram (Fig. 8) than in the SCS (Fig. 4). This is simply because in Fig. 8 we show the maximum amplitude enhancement around the particle, whereas in Fig. 4 we show the SCS, which is related to the field amplitude squared. The maximum amplitude enhancement increases considerably for more complex shapes: whereas it is 6 for the circle, it is about 35 for the triangle.

The differences between the simplex shapes is much more pronounced for the enhancement than for the SCS. For the 20 nm triangle the maximum Raman enhancement exceeds 10^6, and for the square it is still about 10^5, whereas it is below 10^4 for the hexagon, the pentagon and the circle (Fig. 8a).

The field enhancement for the larger simplexes is shown in Fig. 8b, again at a distance of 1 nm from the surface. We observe that the correlation with the SCS diagram is now weaker. The maximum enhancement is comparable to that obtained previously for the smaller particles.

As discussed previously, the local variations of the near-field distribution determine the effective enhancement experienced by a molecule. This is investigated in Fig. 9, where we present the amplitude distribution as a function of the distance from the tip surface for the triangle, square and circle. The data correspond to the main resonance, and two particle sizes are considered.

For the circle, the field amplitude on the surface reaches 6.9 (4.5), for the small (large) circle. At 2 nm from the surface, the amplitude drops to 2.2 (1.7). For the triangle we observe a huge enhancement for both particle sizes: near the surface more than 100 in terms of amplitude (10^8 Raman), and still more than 50 (6×10^6) at a half-nanometer distance (Fig. 9). For the square we obtain a similar behavior, although the amplitudes are about a factor of two smaller (the Raman enhancement being 16 times weaker). The strong

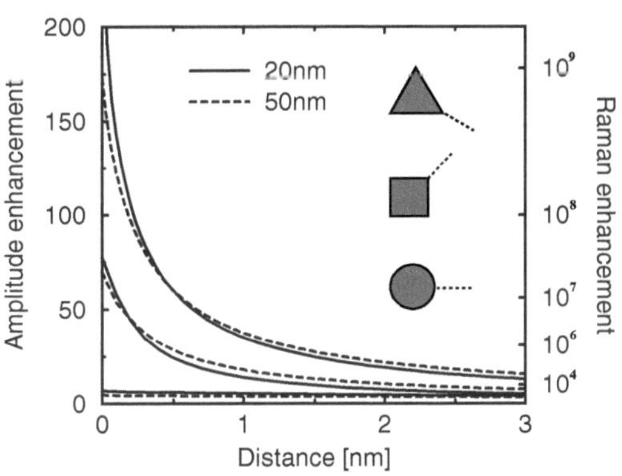

Fig. 9. Enhancement as a function of the distance from the corner (along the *dashed line* in the *inset*), for the main resonance of the 20 nm and 50 nm simplexes. The curves follow the particles order in the inset

field-amplitude variations for such non-regular structures might explain the "hot spots" observed both in SERS experiments and in direct measurements of the locally enhanced field [9,82].

Figure 9 indicates similar results for the maximum amplitude distribution around the 20 nm and the 50 nm simplexes. Although not shown here, this enhancement rapidly decreases for particle sizes above 50 nm [50]. For example, the maximum amplitude enhancement at the corner of a 100 nm triangle is only half that of the 20 or 50 nm particles [54].

Within our model we therefore observe that for a given particle shape the maximum field enhancement increases with decreasing particle sizes down to 50 nm and then remains fairly constant. Note that this maximum enhancement occurs at different wavelengths for different sizes.

Although the local maximum enhancement is similar for particles in the 20–50 nm range, the average over the entire particle may differ with the particle size. This is important for SERS experiments in which large numbers of molecules are used, since the measured Raman signal is proportional to the average Raman enhancement on the surface. This is illustrated in Fig. 5 of [54], where the overall near-field distribution for particles between 10 and 100 nm is shown. We shall now discuss this average enhancement in further details.

In Fig. 10 we show the average enhancement for the 20 nm and 50 nm simplexes. This average value is obtained by taking between 500 and 800 points (depending on the particle shape) distributed regularly around the particle, at a distance of 1 nm from the surface. Note that the average amplitude (Fig. 10a,b) and the average Raman enhancement (Fig. 10c,d) must now be represented on separate graphs.

In Fig. 10a we observe that the average field amplitude is strongly correlated with the SCS (Fig. 4a), both with respect to the wavelength and the width of the different resonances. The average field amplitude enhancement is quite similar for the circle, the hexagon and the pentagon: about 5.5 times the initial amplitude. For the square it is about 7.5 at the corresponding main resonance wavelength, whereas it reaches almost 12 for the triangle.

The average Raman enhancement for the 20 nm simplexes, as shown in Fig. 10c, is less than 10^3 for the circle, becomes larger for the hexagon and the pentagon, and reaches about 10^4 for the square and almost 10^6 for the triangle (always at the corresponding main resonance wavelength). It is important to realize that, due to the rapid variations of the field, the maximum field amplitude on the particle circumference dominates the average Raman enhancement, as the fourth power of the field is taken. This is the reason why the average Raman enhancement is larger than the fourth power of the corresponding average field enhancement (Fig. 10a,c).

Similar results for the 50 nm simplexes are shown in Fig. 10b,d. Comparing Fig. 10a,b, we observe that these larger simplexes produce a smaller average field enhancement than their 20 nm counterparts. This is particularly the

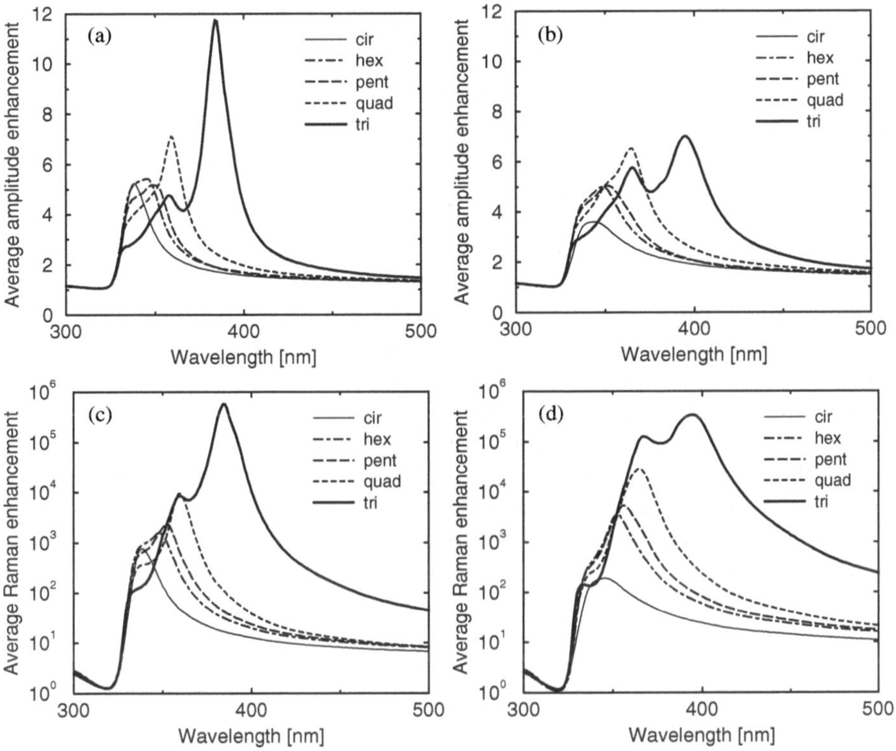

Fig. 10. (**a**),(**b**) Average amplitude enhancement and (**c**),(**d**) average Raman enhancement, computed at a 1 nm distance from the surface of the 20 nm simplexes (**a**),(**c**) and the 50 nm simplexes (**b**),(**d**)

case for the triangular nanowires, where the maximum average enhancement drops from almost 12 to less than 7 times the illumination amplitude. (For the other shapes, the decrease of the maximum average enhancement is less than 15%.) For the average Raman enhancement we also observe a smaller enhancement for the 50 nm circular particle than for its 20 nm counterpart. For the hexagon, pentagon and square particles, the average Raman enhancement is somewhat larger for the 50 nm than for the 20 nm simplexes (Fig. 10c,d). For the triangular particle, the average Raman enhancement at the main resonance is slightly weaker for the 50 nm particle than for the 20 nm one, whereas for the next resonance it is more than ten times stronger. Again, this can be understood with the larger maximum Raman enhancement, as observed in Fig. 8. This larger maximum value outweighs the fact that, for the 50 nm particles, the field amplitude is on average smaller than for the 20 nm particles.

This illustrates the complexity of the interpretation of Raman experiments in which large numbers of molecules are used. The fact that the local

Raman enhancement (Fig. 8) can be much stronger than the average enhancement (Fig. 10) indicates that in such an experiment a very limited number of molecules can contribute the major part of the SERS effect. Let us finally note that the distance between the active Raman site and the surface can also influence the respective magnitude of the local and average enhancements. As a matter of fact, Fig. 9 indicates that at very short distances from the surface, the field is stronger for the 20 nm simplexes. In that case, both the average amplitude and the average Raman enhancements are larger for the 20 nm simplexes.

The large number of resonances observed in non-regular particles opens interesting applications. For example, by tuning the wavelength of the illumination light, it is possible to modify the field distribution in the system and influence also the Raman enhancement, in particular its distance dependence, as illustrated in Fig. 11. The possibility to address different resonances by tuning the illumination wavelength can also prove useful for other specific applications, such as apertureless SNOM [83].

In Fig. 11 we show the field amplitude as a function of the distance from the corner for the three resonances of the 50 nm triangle (see Fig. 4b for the corresponding SCS). We observe a similar enhancement of about 150 close to the surface for the main resonance ($\lambda = 399$ nm) and the resonance at $\lambda = 364$ nm. However, the field decays more rapidly for the latter resonance (Fig. 11). As explained in Sect. 4, this behavior can be related to the polarization charge distribution associated with both resonances.

The extremely important dynamic range observed in Fig. 11 can be useful in an apertureless SNOM experiment, where the tip is vibrated vertically above the sample. It could provide an extremely good signal-to-noise ratio and facilitate the optical feedback of the tip motion.

The other deciding factor that determines the resolution in a SNOM experiment is the lateral field confinement. In [51] we showed that the field distribution obtained with a non-regular nanoparticle remains extremely well confined. At distances between 1 nm and 5 nm from the tip, the full width half-maximum (FWHM) of the lateral field distribution varies between 2 nm

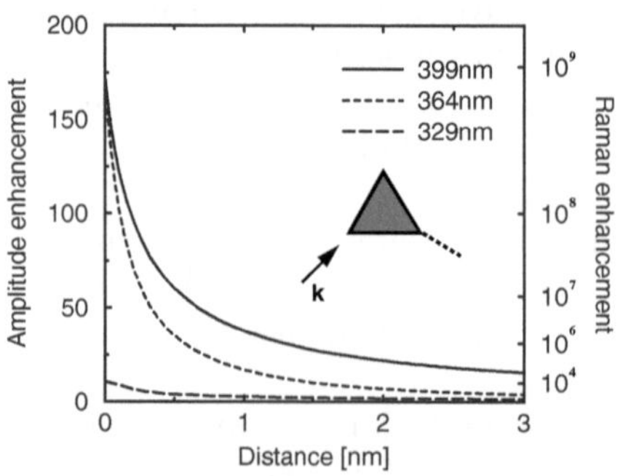

Fig. 11. Field amplitude enhancement as a function of the distance from the corner of the 50 nm triangle along the *dashed line* in the *inset*, for three different resonance wavelengths (SCS in Fig. 4b)

and 10 nm [51]. In the close vicinity of the particle, the field is confined over dimensions much smaller than the particle size, as illustrated in Fig. 7a,b.

6 Influence of the Model Permittivity and of the Background

The previous results emphasize how complex the study of plasmon resonances can be. In particular, the comparison between different particle shapes is quite delicate, as the corresponding resonances occur at specific frequencies, with different permittivities. In particular, ε'' – which accounts for the absorption – strongly influences the resonance width and the near-field enhancement. To clarify this point, we concentrate in this section on the triangular particle investigated in Fig. 5 and study the influence of the absorption on the SCS and near-field enhancement.

The definition of the permittivity for a nanoscopic particle is not trivial. For particles smaller than 2 nm, the classical description breaks down and jellium or quantum-chemistry models must be used [65,67]. For the particles in the 10–50 nm range considered here, a local permittivity can still be used to describe the metal, but the bulk value must be modified as the particle size becomes comparable to the bulk electron mean free path (the average path of a conduction electron between two scattering processes). In that case, electron scattering at the particle surface becomes a dominant effect. It reduces the electron mean free path, which in turn leads to an increase of the imaginary part of the permittivity. This has been shown experimentally for spherical particles by *Kreibig* et al. [59,63,84]. They showed that for particle sizes below 40 nm the resonance width becomes broader, and in particular they found that near the plasma resonance frequency the imaginary part can be well described by the relation:

$$\varepsilon'' = 0.23 + \frac{2.64}{a}, \tag{2}$$

where a is the particle size in nanometers [58,84].

More recently, the electron dynamics in silver and copper particles in the 5–10 nm range was investigated using femtosecond laser pulses [64]. It was found that for a 6.5 nm particle size the resonance width is between 3 and 4 times larger than expected from the bulk permittivity, thereby indicating a larger absorption. The resonance was also slightly blue shifted, indicating that the real part of the permittivity can also be affected for such a small particle size [64]. Similar results have been obtained for silver particles in the 3–13 nm range by another group [85]. For spherical particles in the 20 nm range investigated here, only the imaginary part of the permittivity seems to be affected by the particle size [58,59].

No experimental data are available for the modification of the dispersion relation in non-spherical particles of a given shape. One can, however, expect that corners produce additional electron scattering, thereby locally increasing the absorption. Following (2), and to gain insights in the influence of the absorption on the particle response, we investigate three different absorptions.

Figure 12 shows the SCS and the near-field distribution for the same right-angled triangular particle as in Fig. 5, using three different values for the imaginary part ε'' of the permittivity. The original values of Johnson and Christy, as measured for the bulk, produce four well-resolved resonances. The main resonance is quite narrow (FWHM=16 nm). This resonance broadens when the absorption increases: it is FWHM = 28 nm for $\varepsilon' + i2\,\varepsilon''$ and FWHM = 52 nm for $\varepsilon' + i4\,\varepsilon''$ (Fig. 12a).

Further, the absorption has a strong influence on the higher-order resonances, which progressively merge into one broad resonance, as the imaginary part of the permittivity increases. Note in Fig. 12a that the resonances are not shifted, since only the imaginary part of the permittivity changes (the plasmon resonance condition being solely determined by the real part of the permittivity).

The influence of the damping factor on the near-field enhancement is investigated in Fig. 12b, for the same particle. This effect is quite dramatic, as the field amplitude for the main resonance at 0.5 nm from the surface drops from 112 ($\varepsilon' + i\,\varepsilon''$) to 65 ($\varepsilon' + i\,2\varepsilon''$) and 36 ($\varepsilon' + i\,4\varepsilon''$). This influence of the particle absorption remains visible at larger distances from the surface (Fig. 12b). A similar behavior is observed for higher-order resonances, although in that case the field dies out rapidly when one moves away from the surface [53].

Let us finally study the influence of the particle background on the resonance spectrum. This is evidently important for experiments in solutions, as well as for many experiments on surfaces, as a thin water layer is often

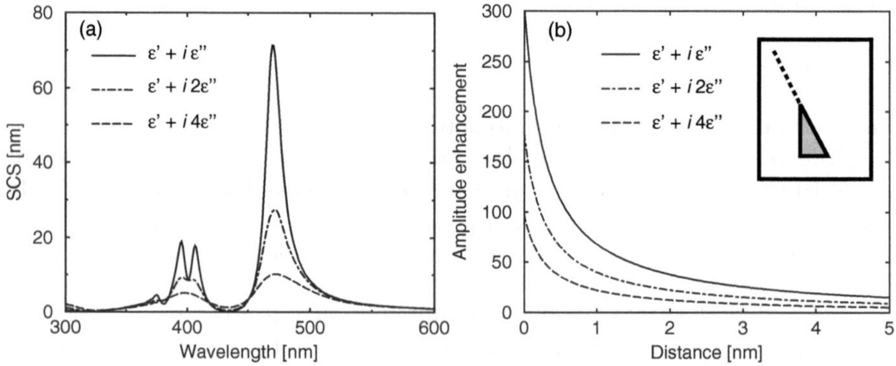

Fig. 12. (**a**) SCS for a triangular particle using different permittivities with increasing imaginary parts ε'' (absorption factor). (**b**) Field amplitude enhancement as a function of the distance from the corner (see *inset*), for the main resonance of the particle in (**a**) ($\lambda = 470\,\text{nm}$)

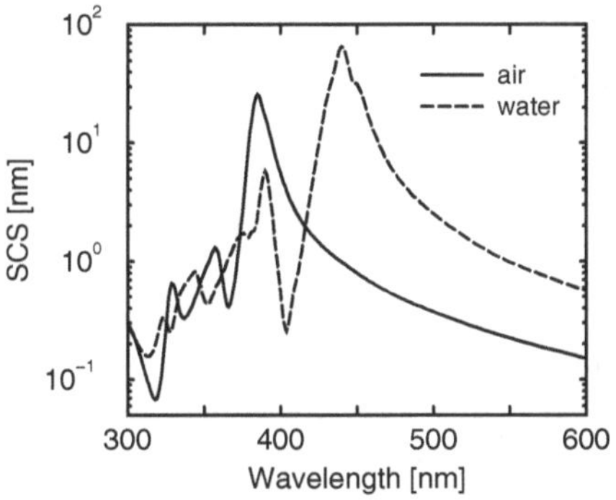

Fig. 13. SCS for a triangular particle (side 20 nm) in air and in water backgrounds

present in that case. Figure 13 shows the SCS for a triangular particle in air ($\epsilon_B = 1$) and in water ($\epsilon_B = 1.78$, water dispersion is negligible in the visible range and the permittivity can be assumed constant [86]).

We observe that for a higher background permittivity the resonance spectrum is red-shifted and broadened, very similarly to the case where the particle size increases (e.g. compare with Fig. 4a,b). As a matter of fact, for larger background permittivity, the particle simply becomes larger with respect to the effective illumination wavelength in the background; retardation – which is responsible for the size effect – then comes into play for smaller particle sizes.

7 Conclusions and Outlook

Our results shed new light on the very old problem of the relation between a nanoparticle shape and its plasmon resonance spectrum. For the first time it is clearly demonstrated for particles with a non-regular shape, that the number of resonances strongly depends on the form of the particle. The higher the particle symmetry, the simpler is its spectrum (e.g. a small cylindrical particle exhibits only one resonance, whereas a square has two and an equilateral triangle at least three distinct resonances). Several additional resonances are observed for right-angled triangular particles with a high perpendicular-to-base ratio.

These complex scattering cross-sections are also associated with a dramatic near-field enhancement in the close vicinity of the particle. We found the strongest enhancement for particles with dimensions smaller than 50 nm. Right-angled triangular particles produce a field amplitude many hundred times stronger than the incident field at short distances from the surface. This dramatic near-field enhancement corresponds to a huge Raman enhancement, similar to that required by recent SERS experiments, where single molecule detection was possible [8,9,87].

The strong localization of the near-field at specific positions around the metallic particle, as well as its rapid decay when one moves further away from the particle surface, can also explain the "hot-spots" observed in SERS experiments, where specific sites appear to be particularly active. This is also confirmed by our results on the average Raman enhancement, which indicate that a limited number of molecules can contribute the major part of the SERS signal, even in experiments in which large numbers of molecules are used.

The topology of the field distribution inside and in the vicinity of the particle can be easily understood in terms of the polarization charges. To each plasmon resonance corresponds a specific charge distribution, which produces a particular field. The field distribution associated with each plasmon resonance is different, in particular its decay rate as one moves away from the particle surface. This can be useful for near-field optical microscopy and other local probe techniques that rely on strongly localized electromagnetic fields.

The nanowires investigated here can now be fabricated in a controlled manner using different nanofabrication techniques. A direct experimental verification of our results is therefore possible. The next challenge however, is a fully three-dimensional (3D) model, which can handle arbitrary shape particles. Our 2D results indicate that at the plasmon resonance the field distribution strength depends on the confinement of the polarization charges at specific locations in the particle. For nanowires, this confinement can take place in the two transverse dimensions. Therefore, for 3D particles, where the charges can be confined in all three directions, we expect an even stronger enhancement.

Acknowledgements

It is a pleasure to acknowledge J. P. Kottmann who developed the finite elements technique and calculated most of the configurations presented here in the course of his Ph.D. I am also indebted to D. R. Smith and S. Schultz for many stimulating discussions.

References

1. M. Kerker: Founding fathers of light scattering and surface-enhanced Raman scattering, Appl. Opt. **30**, 4699 (1991) 183
2. R. Jin, Y. W. Cao, C. A. Mirkin, K. L. Kelly, G. C. Schatz, J. G. Zheng: Photoinduced conversion of silver nanospheres to nanoprism, Science **294**, 1901 (2001) 184
3. J. J. Mock, M. Barbic, D. R. Smith, D. A. Schultz, S. Schultz: Shape effects in plasmon resonance of individual colloidal silver nanoparticles, J. Chem. Phys. **116**, 6755 (2002) 184
4. M. Fleischmann, P. J. Hendra, A. J. McQuillan: Raman-spectra of pyridine adsorbed at a silver electrode, Chem. Phys. Lett. **26**, 163 (1974) 184
5. D.L. Jeanmaire, R.P. van Duyne: Surface Raman spectroelectrochemistry. 1. Heterocyclic, aromatic, and aliphatic-amines adsorbed on anodized silver electrodes, J. Electroanal. Chem. **84**, 1 (1977) 184

6. H. Metiu: Surface enhanced spectroscopy, Prog. Surf. Sci. **17**, 153 (1984) 184
7. M. Moskovits: Surface-enhanced spectroscopy, Rev. Mod. Phys. **57**, 783 (1985) 184, 196
8. K. Kneipp, Y. Wang, H. Kneipp, L. T. Perelman, I. Itzkan, R. R. Dasari, M. S. Feld: Single molecule detection using surface-enhanced Raman scattering, Phys. Rev. Lett. **78**, 1667 (1997) 184, 203
9. S. Nie, S. R. Emory: Probing single molecules and single nanoparticles by surface-enhanced Ramsn scattering, Science **275**, 1102 (1997) 184, 198, 203
10. H. Xu, E. J. Bjerneld, M. Käll, L. Börjesson: Spectroscopy of single Hemoglobin molecules by surface enhanced Raman scattering, Phys. Rev. Lett. **83**, 4357 (1999) 184, 188
11. A. Otto, I. Mrozek, H. Grabhorn, W. Akemann: Surface-enhanced Raman scattering, J. Phys. C **4**, 1143 (1992) 184, 197
12. P. Kambhampati, C. M. Child, M. C. Foster, A. Campion: On the chemical mechanism of surface enhanced Raman scattering: Experiment and theory, J. Chem. Phys. **108**, 5013 (1998) 184, 197
13. J. R. Krenn, A. Dereux, J. C. Weeber, E. Bourillot, Y. Lacroute, J. P. Goudonnet, G. Schider, W. Gotschy, A. Leitner, F. R. Aussenegg, C. Girard: Squeezing the optical near–field by plasmon coupling of metallic nanoparticles, Phys. Rev. Lett. **82**, 2590 (1999) 184
14. J.-C. Weeber, A. Dereux, C. Girard, J. R. Krenn, J.-P. Goudonnet: Plasmon polaritons of metallic nanowires for controlling submicron propagation of light, Phys. Rev. B **60**, 9061 (1999) 184
15. T. Yatsui, M. Kourogi, M. Ohtsu: Plasmon waveguide for optical far/near-field conversion, Appl. Phys. Lett. **79**, 4583 (2001) 184
16. J. Tominaga, T. Nakano, N. Atoda: Super-resolution structure for optical data storage by near-field optics, Proc. SPIE **3467**, 282 (1999) 184
17. L. Men, J. Tominaga, H. Fuji, Q. Chen, N. Atoda: High-density optical data storage using scattering-mode super-resolution near-field structure, Proc. SPIE **4085**, 204 (2001) 184
18. J. Tominaga, C. Mihalcea, D. Büchel, H. Fukuda, T. Nakano, N. Atoda, H. Fuji, T. Kikukawa: Local plasmon photonic transistor, Appl. Phys. Lett. **78**, 2417 (2001) 184
19. K. Bromann, C. Félix, H. Brune, W. Harbich, R. Monot, J. Buttet, K. Kern: Controlled deposition of size-selected Silver nanoclusters, Science **274**, 956 (1996) 185
20. D. M. Kolb, R. Ullmann, T. Will: Nanofabrication of small copper clusters on gold(111) electrodes by a scanning tunneling microscope, Science **275**, 1097 (1997) 185
21. Y.-Y. Yu, S.-S. Chang, C.-L. Lee, C. R. C. Wang: Gold nanorods: Electrochemical synthesis and optical properties, J. Phys. Chem. B **101**, 6661 (1997) 185
22. K. Abe, T. Hanada, Y. Yoshida, N. Tanigaki, H. Takiguchi, H. Nagasawa, M. Nakamoto, T. Yamaguchi, K. Yase: Two-dimensional array of silver nanoparticles, Thin Solid Films **327–329**, 524 (1997) 185
23. D. Y. Petrovykh, F. J. Himpsel, T. Jung: Width distribution of nanowires grown by step decoration, Surf. Science **407**, 189 (1998) 185
24. G. L. Che, B. B. Lakshmi, E. R. Fisher, C. R. Martin: Carbon nanotubule membranes for electrochemical energy storage and production, Nature **393**, 346 (1998) 185

25. J. Viereck, F. Stietz, M. Stuke, T. Wenzel, F. Träger: The role of surface defects in laser-induced thermal desorption from metal surfaces, Surf. Sci. **383**, 749 (1997) 185
26. J. Bosbach: Laser-based method for fabricating monodispersive metallic nanoparticles, Appl. Phys. Lett. **74**, 2605 (1999) 185
27. I. Utke, P. Hoffmann, B. Dwir, E. Kapon, P. Doppelt: Focused electron beam induced deposition of gold, J. Vac. Sci. Technol. B **18**, 3168 (2000) 185
28. A. P. Li, F. Müller, U. Gösele: Polycrystalline and monocrystalline pore arrays with large interpore distance in anodic alumina, Electrochem. Solid-State Lett. **3**, 131 (2000) 185
29. D.-S. Wang, H. Chew, M. Kerker: Enhanced Raman scattering at the surface of a spherical particle, Appl. Opt. **19**, 2256 (1980) 185
30. M. Kerker, D.-S. Wang, H. Chew: Surface enhanced Raman scattering (SERS) by molecules adsorbed at spherical particles: errata, Appl. Opt. **19**, 4159 (1980) 185
31. P. K. Aravind, A. Nitzan, H. Metiu: The interaction between electromagnetic resonances and its role in spectroscopic studies of molecules adsorbed on colloidal particles or metal spheres, Surf. Sci. **110**, 189 (1981) 185
32. P. W. Barber, R. K. Chang, H. Massoudi: Electrodynamic calculations of the surface-enhanced electric intensities on large Ag spheroids, Phys. Rev. B **27**, 7251 (1983) 185
33. M. Inoue, K. Ohtaka: Enhanced Raman scattering by two-dimensional array of polarizable spheres, J. Phys. Soc. Jpn. **52**, 1457 (1983) 185
34. K. T. Carron, W. Fluhr, M. Meier, A. Wokaun, H. W. Lehmann: Resonances of two-dimensional particle gratings in surface-enhanced Raman scattering, J. Opt. Soc. Am. B **3**, 430 (1986) 185
35. R. Rojas, F. Claro: Theory of surface enhanced Raman scattering in colloids, J. Chem. Phys. **98**, 998 (1993) 185
36. A. I. Vanin: Surface-amplified Raman scattering of light by molecules adsorbed on groups of spherical particles, J. Appl. Spectrosc. **62**, 32 (1995) 185
37. K.-P. Charlé, L. König, S. Nepijko, I. Rabin, W. Schulze: The surface plasmon resonance in free and embedded Ag-clusters in the size range $1,5\,\mathrm{nm} < \mathrm{D} < 30\,\mathrm{nm}$, Cryst. Res. Technol. **33**, 1085 (1998) 185, 187
38. J. Gersten, A. Nitzan: Electromagnetic theory of enhanced Raman scattering by molecules adsorbed on rough surfaces, J. Chem. Phys. **73**, 3023 (1980) 185
39. M. Weber, D. L. Mills: Interaction of electromagnetic waves with periodic gratings: Enhanced fields and the reflectivity, Phys. Rev. B **27**, 2698 (1983) 185
40. F. J. García-Vidal, J. B. Pendry: Electromagnetic interactions with rough metal surfaces, Prog. Surf. Sci. **50**, 55 (1995) 185
41. S. J. Oldenburg, R. D. Averitt, S. L. Westcott, N. J. Halas: Nanoengineering of optical resonances, Chem. Phys. Lett. **288**, 243 (1998) 185
42. F. J. García-Vidal, J. B. Pendry: Collective theory for surface enhanced Raman scattering, Phys. Rev. Lett. **77**, 1163 (1996) 185
43. M. Kahl, E. Voges, S. Kostrewa, C. Viets, W. Hill: Periodically structured metallic substrates for SERS, Sens. Actuators B **51**, 285 (1998) 185
44. M. Kahl, E. Voges: Analysis of plasmon resonance and surface-enhanced Raman scattering on periodic silver structures, Phys. Rev. B **61**, 14078 (2000) 185
45. R. Fuchs: Theory of the optical properties of ionic crystal cubes, Phys. Rev. B **11**, 1732 (1975) 185

46. W.-H. Yang, G. C. Schatz, R. P. van Duyne: Discrete dipole approximation for calculating extinction and Raman intensities for small particles with arbitrary shape, J. Chem. Phys. **103**, 869 (1995) 185, 188
47. T. R. Jensen, G. C. Schatz, R. P. van Duyne: Nanosphere lithography: surface plasom resonance spectrum of a periodic array of silver nanoparticles by ultraviolett-visible extinction spectroscopy and electrodynamic modeling, J. Phys. Chem. B **103**, 2394 (1999) 185, 188
48. N. Félidj, J. Aubard, G. Lévi: Discrete dipole approximation for ultraviolet-visible extinction spectra simulation of silver and gold colloids, J. Chem. Phys. **111**, 1195 (1999) 185, 188
49. J. P. Kottmann, O. J. F. Martin: Accurate solution of the volume integral equation for high permittivity scatterers, IEEE Trans. Antennas Propag. **48**, 1719 (2000) 185, 189, 190
50. J. P. Kottmann, O. J. F. Martin, D. R. Smith, S. Schultz: Dramatic localized electromagnetic enhancement in plasmon resonant nanowires, Chem. Phys. Lett. **341**, 1 (2001) 185, 186, 198
51. J. P. Kottmann, O. J. F. Martin, D. R. Smith, S. Schultz: Non–regularly shaped plasmon resonant nanoparticle as localized light source for near–field microscopy, J. Microscopy **202**, 60 (2001) 185, 195, 200, 201
52. J. P. Kottmann, O. J. F. Martin, D. R. Smith, S. Schultz: Plasmon resonances of silver nanowires with a non–regular cross–section, Phys. Rev. B **64**, 235 402 (2001) 185, 190
53. J. P. Kottmann, O. J. F. Martin: Influence of the cross section and the permittivity on the plasmon-resonance spectrum of silver nanowires, Appl. Phys. B **73**, 299 (2001) 185, 202
54. J. P. Kottmann, O. J. F. Martin, D. R. Smith, S. Schultz: Spectral response of Silver nanoparticles, Optics Express **6**, 213 (2000) 185, 192, 195, 198
55. J. P. Kottmann, O. J. F. Martin, D. R. Smith, S. Schultz: Field polarization and polarization charge distributions in plasmon resonant particles, New J. Phys. **2**, 27.1 (2000) 185, 192, 193, 195
56. J. P. Kottmann, O. J. F. Martin: Plasmon resonant coupling in metallic nanowires, Optics Express **8**, 655 (2001) 185, 195
57. J. P. Kottmann, O. J. F. Martin: Retardation–induced plasmon resonances in coupled nanoparticles, Opt. Lett. **26**, 1096 (2001) 185, 195
58. C. F. Bohren, D. R. Huffman: *Absorption and Scattering of Light by Small Particles* (Wiley, New York 1983) 186, 188, 189, 201
59. U. Kreibig, M. Vollmer: *Optical Poperties of Metal Clusters*, Springer Ser. Mater. Sci. **25** (Springer-Verlag, Berlin, Heidelberg 1995) 186, 187, 201
60. P. B. Johnson, R. W. Christy: Optical constants of the noble metals, Phys. Rev. B **6**, 4370 (1972) 186, 187
61. R. Ruppin: Spherical and cylindrical surface polaritons in solids, in A. D. Boardman (Ed.): *Electromagnetic Surface Modes* (Wiley, Chichester 1982) 186
62. U. Kreibig, C. von Fragstein: The limitation of electron mean free path in small silver particles, Z. Physik **224**, 307 (1969) 187
63. L. Genzel, T. P. Martin, U. Kreibig: Dielectric function and plasma resonances of small metal particles, Z. Physik B **21**, 339 (1975) 187, 201
64. J.-Y. Bigot, V. Halté, J. C. Merle, A. Daunois: Electron dynamics in metallic nanoparticles, Chem. Phys. **251**, 181 (2000) 187, 201
65. M. Brack: The physics of simple metal clusters: self-consistent jellium model and semiclassical approaches, Rev. Mod. Phys. **65**, 677 (1993) 187, 201

66. V. V. Kresin: Collective resonances and response properties of electrons in metal clusters, Phys. Rep. **220**, 1 (1992) 187
67. V. Bonacic-Koutecky, P. Fantucci, J. Koutecky: Quantum-chemistry of small clusters of elements of group-IA, group-IB, and group-IIA: Fundamental concepts, predictions and interpretation of experiments, Chem. Rev. **91**, 1035 (1991) 187, 201
68. O. J. F. Martin, N. B. Piller: Electromagnetic scattering in polarizable backgrounds, Phys. Rev. E **58**, 3909 (1998) 188
69. M. Paulus, P. Gay-Balmaz, O. J. F. Martin: Accurate and efficient computation of the Green's tensor for stratified media, Phys. Rev. E **62**, 5797 (2000) 188
70. M. Paulus, O. J. F. Martin: Light propagation and scattering in stratified media: A Green's tensor approach, J. Opt. Soc. Am. A **18**, 854 (2001) 188
71. A. Taflove: *Computational Electrodynamics, the FDTD Method* (Artech House, Boston 1995) 188
72. B. T. Draine, P. J. Flatau: Discrete–dipole approximation for scattering calculations, J. Opt. Soc. Am. A **11**, 1491 (1994) 188
73. P. J. Flatau: Improvements in the discrete–dipole approximation method of computing scattering and absorption, Opt. Lett. **22**, 1205 (1997) 188
74. M. I. Stockmann, V. M. Shalaev, M. Moskovits, R. Botet, T. F. George: Enhanced Raman scattering by fractal clusters: Scale-invariant theory, Phys. Rev. B **46**, 2821 (1992) 188
75. N. B. Piller, O. J. F. Martin: Increasing the performances of the coupled–dipole approximation: A spectral approach, IEEE Trans. Antennas Propag. **46**, 1126 (1998) 188
76. J. van Bladel: *Singular Electromagnetic Fields and Sources* (Clarendon, Oxford 1991) 190, 195
77. L. D. Landau, E. M. Lifshitz, L. P. Pitaevskii: *Electrodynamics of continuous media*, Vol. 8 of *Landau and Lifshitz course on theoretical physics*, 2nd. ed. (Butterworth Heinemann, Oxford 1998) 193
78. B. Hecht, H. Bielefeld, Y. Inouye, D. W. Pohl, L. Novotny: Facts and artifacts in near-field optical microscopy, J. Appl. Phys. **81**, 2492 (1997) 195
79. R. P. van Duyne: private communication 196
80. M. D. Malinsky, L. Kelly, G. C. Schatz, R. P. van Duyne: Chain length dependance and sensing capabilities of the localized surface plasmon resonance of silver nanoparticles chemically modified with Alkanethiol self-assembled monolayers, J. Am. Chem. Soc. **123**, 1471 (2001) 196
81. A. Campion, P. Kambhampati: Surface-enhanced Raman scattering, Chem. Soc. Rev. **27**, 241 (1998) 197
82. P. Gadenne, X. Quelin, S. Ducourtieux, S. Gresillon, L. Aigouy, J.-C. Rivoal, V. Shalaev, A. Sarychev: Direct observation of locally enhanced electromagnetic fields, Physica B **279**, 52 (2000) 198
83. Y. C. Martin, H. K. Wickramasinghe: Resolution test for apertureless near-field optical microscopy, J. Appl. Phys. **91**, 3363 (2002) 200
84. U. Kreibig, C. von Fragstein: Electronic properties of small silver particles: the optical constants and their temperature dependence, J. Phys. F. **4**, 999 (1974) 201
85. N. D. Fatti, F. Vallée, C. Flytzanis, Y. Hamanaka, A. Nakamura: Electron dynamics and surface plasmon resonance nonlinearities in metal nanoparticles, Chem. Phys. **251**, 215 (2000) 201

86. Optical Society of America, in M. Bass (Ed.): *Handbook of Optics*, Vol. II., 2nd ed. (McGraw Hill, New York 1995) 203
87. H. Xu, J. Aizpurua, M. Käll, P. Apell: Electromagnetic contributions to single-molecule sensitivity in surface-enhanced Raman scattering, Phys. Rev. E **62**, 4318 (2000) 203

Index

Topics in Applied Physics

71 **The Monte Carlo Method in Condensed Matter Physics**
By K. Binder (Ed.) 2nd edn. 1995. 83 figs. XX, 418 pages

72 **Glassy Metals III**
Amorphization Techniques, Catalysis, Electronic and Ionic Structure
By H. Beck and H.-J. Güntherodt (Eds.) 1994. 145 figs. XI, 259 pages

73 **Hydrogen in Metals III**
Properties and Applications
By H. Wipf (Ed.) 1997. 117 figs. XV, 348 pages

74 **Millimeter and Submillimeter Wave Spectroscopy of Solids**
By G. Grüner (Ed.) 1998. 173 figs. XI, 286 pages

75 **Light Scattering in Solids VII**
Christal-Field and Magnetic Excitations
By M. Cardona and G. Güntherodt (Eds.) 1999. 96 figs. X, 310 pages

76 **Light Scattering in Solids VIII**
C60, Semiconductor Surfaces, Coherent Phonons
By M. Cardona and G. Güntherodt (Eds.) 1999. 86 figs. XII, 228 pages

77 **Photomechanics**
By P. K. Rastogi (Ed.) 2000, 314 Figs. XVI, 472 pages

78 **High-Power Diode Lasers**
By R. Diehl (Ed.) 2000, 260 Figs. XIV, 416 pages

79 **Frequency Measurement and Control**
Advanced Techniques and Future Trends
By A. N. Luiten (Ed.) 2001, 169 Figs. XIV, 394 pages

80 **Carbon Nanotubes**
Synthesis, Structure, Properties, and Applications
By M. S. Dresselhaus, G. Dresselhaus, Ph. Avouris (Eds.) 2001, 235 Figs. XVI, 448 pages

81 **Near-Field Optics and Surface Plasmon Polaritons**
By S. Kawata (Ed.) 2001, 136 Figs. X, 210 pages

82 **Optical Properties of Nanostructured Random Media**
By Vladimir M. Shalaev (Ed.) 2002, 185 Figs. XIV, 450 pages

83 **Spin Dynamics in Confined Magnetic Structures I**
By B. Hillebrands and K. Ounadjela (Eds.) 2002, 166 Figs. XVI, 336 pages

84 **Imaing of Complex Media with Acoustic and Seismic Waves**
By M. Fink, W. A. Kuperman, J.-P. Montagner, A. Tourin (Eds.) 2002, 162 Figs. XII, 336 pages

85 **Solid–Liquid Interfaces**
Macroscopic Phenomena – Microscopic Understanding
By K. Wandelt and S. Thurgate (Eds.) 2003, 228 Figs. XVIII, 444 pages

86 **Infrared Holography for Optical Communications**
Techniques, Materials, and Devices
By P. Boffi, D. Piccinin, M. C. Ubaldi (Eds.) 2003, 90 Figs. XII, 182 pages

87 **Spin Dynamics in Confined Magnetic Structures II**
By B. Hillebrands and K. Ounadjela (Eds.) 2003, 179 Figs. XVI, 321 pages

88 **Optical Nanotechnologies**
The Manipulation of Surface and Local Plasmons
By J. Tominaga and D. P. Tsai (Eds.) 2003, XII Figs. 168, 212 pages

MIX
Papier aus verantwortungsvollen Quellen
Paper from responsible sources
FSC® C105338

If you have any concerns about our products,
you can contact us on
ProductSafety@springernature.com

In case Publisher is established outside the EU,
the EU authorized representative is:
Springer Nature Customer Service Center GmbH
Europaplatz 3, 69115 Heidelberg, Germany

Printed by Libri Plureos GmbH
in Hamburg, Germany